Window on the Forth

Window on the Forth

Malcolm Archibald

Contents

FOR CATHY

Introduction

For centuries, the Firth of Forth has been the maritime gateway into Scotland, a great bite in the North Sea Coast that invites trade and allows entry into the heart of the nation. There is a tremendous beauty here, while every island, neuk and cove shelters a wealth of history, stories and legend. As a North Sea trading area, the Forth has welcomed vessels from all across the globe, from the pragmatic galleys of Rome to cogs from France and fat-bellied Dutch busses, and from the blubber-smeared whaling ships of Leith, Bo'ness and Fife to the giant tankers that berth off Hound Point.

In the distant past, people termed the Forth as Scotswater or the Scottish Sea, and salt-stained ships from Fife passed over to Norway and the Netherlands. If it was an international highway, it was also part of the transport network of old Scotland, for muscular oarsmen propelled their ferries on half a score of routes, carrying queens and commoners, outlaws and clergy from one side of the Forth to the other. There were many holy men here, for Inchcolm was the East's answer to Iona, and the Forth was also on the direct pilgrim route to the shrines at St. Andrews. Not all the pilgrims and clergy were as holy as they might have been, for was it not a crew of monks who allegedly tipped Lord Mortimer into the Deep that bears his name? In time the construction of the great bridges relinquished the ferries to history and now trains, lorries and cars drive high above the blue firth.

Fishermen have long frequented this water. Fishing villages decorate the shores: Newhaven, Cockenzie; Dunbar, Anstruther, Pittenweem and Crail are only a few of the Forth-side communities that have sent generation after generation of men out to fish the deep waters, often with tragic results. The sea is not always a kind mistress.

Not all the visiting vessels were friendly. On many occasions, hostile ships entered the Forth: English intending piracy or conquest, Dutchmen looking to bombard Burntisland, American privateers searching for plunder and on one occasion a French invasion fleet that anchored off Crail. In the World Wars of the twentieth century, German submarines probed the Forth defences and men watched the waters for hostile periscopes and the sinister growl of enemy aircraft. Just as alarming for the early Forth mariners were the Royal Navy press gangs, who could legally kidnap a man so he could disappear over the horizon for years at a time. Of course, not all Forth seamen were innocent; one of the most famous fictional survival tales of all time, *Robinson Crusoe*, was based on the exploits of a Forth privateer. Not surprisingly, the Firth sent her seagoing sons all around the world and when possibly the greatest clipper tea race of the nineteenth century took place; both leading ships had a Fife connection.

Because of its history, character and atmosphere, the Firth of Forth has bred some of the finest seamen in the world. This book will open a window onto just a few of their stories, and speckle them with short histories of the islands that decorate the Firth.

Structure

In a book of this type, where geography and history combine with folklore and legend, the structure presented significant problems. A straightforward chronological history might have been a simpler option, but would have necessarily broken up the themes that are the essence of the story. My idea was to present a series of easily digestible articles, each of which illustrates one aspect, or one chapter of history, or one location, of the Firth of Forth.

With that criterion in mind, I decided on a simple structure that combined themes and rough chronology. Each chapter looks at one aspect of the Forth in the following order:

Chapter one: Fleets in the Forth, 15th and 16th century
Chapter two: the Outer Islands, history and geography
Chapter three: The Emigrant Fleets, the Darien Scheme, 17th century
Chapter four: The Real Robinson Crusoe, 17th century
Chapter five: Jacobites in the Forth, 18th century
Chapter six: The French Wars, 18th and 19th centuries
Chapter seven: The Greenlandmen, 18th and 19th centuries
Chapter eight: The Inner Islands, history and geography
Chapter nine: Fisherfolk, 19th century
Chapter ten: The Great Tea Race, 19th century
Chapter eleven: Magnificent Men in their Flying Machines, 19th and 20th century
Chapter twelve: Crossing the Pond, 19th century
Chapter thirteen: The German Wars, 20th century
Chapter fourteen: Crossing the Forth, 19th century
Chapter fifteen: The Bridges, 19th and 20th century

The material is anecdotal rather than academic, and the book is intended to entertain and perhaps introduce some of the fascinating events that occurred here, and the people who have called the Firth of Forth their home.

I hope you enjoy reading it as much as I enjoyed the research and writing of it.

Chapter One

Fleets in the Forth

The fifteenth century saw the culmination of dramatic improvements in North European shipbuilding technology. For centuries, men had built clinker-fashion, with one plank overlapping the next, but now ships could be constructed with the hull planks meeting flush. The number of masts that vessels carried had also multiplied from one to three, with square sails on the foremast and mainmast, and a lateen sail on the mizzen. The new type of vessel thus created was known as a caravel or carvel; she was lighter in weight, and broader of beam than the previous clinker built craft, and she pointed her bowsprit toward a new, worldwide horizon.

Perhaps her high poop and clumsy forecastle would make her ungainly to modern eyes, but caravels were the cutting edge of maritime technology in their day. As far back as the thirteenth century, Portuguese fishing *caravelas* had pushed out into the broad Atlantic, but not until the middle of the fifteenth did the caravels of Iberia, equipped with the newly created compass, plough uncharted sea roads to find unknown peoples, novel cultures and distant continents. It was the carvel that first opened up the oceans to European navigators and nothing would ever be the same again.

Brightest in Iberia, light from this new nautical dawn filtered to the remotest coasts of Europe and eventually to Scotland. Blessed with a long seacoast, Scotland was a maritime nation seemingly without ambition; her sea masters followed established trade routes to Europe and her fishermen rarely left the stormy security of her own shores. After generations of trade with the Low Countries, the Scottish colony in Bruges had its own church in which the congregation erected an altar to St Ninian, Scotland's first saint, and supported a chaplain to say masses for the welfare of seamen's souls. In 1467 the parliament of James III established tolls on freight money to support this cause, with every ship with a cargo of more than one hundred tons paying the Scots chaplain at Bruges one sack of goods, while ships with less than one hundred tons parted with half a sack. Such seemed the extent of Scotland's nautical aspiration.

However Scotland is a land of surprises, and once her seamen learned of the carvels, they quickly adopted them. Scotland moved forward in maritime skills. In 1448 Alexander Wallace of Leith skippered a carvel to East Anglia and proved the worth of his ship when he captured a vessel from Hastings. In the 1450s Bishop Kennedy of St Andrews built *Salvador*, described as a 'ship the biggest that had been seen to sail upon the ocean'. Around 1455 an inventory of the port of Sluys numbered seventy-two ships, of which six were Scottish and the largest, at 500 tons, belonged to the Bishop of St Andrews. It is likely that this ship was *Salvador*; a forerunner of the shipbuilding skills that were to astonish the world.

But even bishops must bow to the gods of the ocean, and in 1474 a storm drove *Salvador* onto the shore of Northumberland. The locals, delighted with the sea's bounty, swarmed to the beach, full of the joys of pillage. If the Abbot of Inchcolm, passenger on board, expected his Holy Orders to protect him from the English habit of kidnap and ransom, he was disappointed, but other travellers and other vessels were equally at risk in that era of habitual piracy.

Two years later King Edward IV of England was apologising to King James III of Scots and offering restitution for Scottish losses to the frequent English raiders. Two of the known instances saw the English Lord Grey pirate the trading vessel of Sir John Colquhoun of Luss, while a ship belonging to the Duke of Gloucester had captured the king's own *Yellow Carvel*.

Aware of the problems faced by his seamen, James III ordered two galleys to counter English piracy. Despite the high price of £300 paid for these vessels, Scotland required more to prevent English peace-time raiding. The English appeared to relish the challenge for in the summer of 1481 they sailed to the Forth and pirated eight vessels. The Scots did not sit idly by, however, and gathered a fleet from Leith to remove this infestation on the firth. There are few surviving details, but the English appear to have been chased away after burning Blackness. Perhaps this raid was the final catalyst for a Scottish nautical revival, for Renaissance Scotland was about to produce a generation of seamen whose name rustle down the centuries like the mainsail of a carvel hoisted for a long ocean voyage.

Admiral of the Scottish Sea

There was William Todrig; Captain Merrimonth and Captain Brownhill; Andrew, John and Robert Barton; William Paterson and Andrew Wood. Between them, this band of Forth seamen carried the Saltire to high prominence from the Baltic to the Azores and for a few golden years, neither pirate nor English raider could safely dare the seas off Scotland. The most prominent, or the best remembered, of these men were the Barton family of Leith and Sir Andrew Wood of Kirkton of Largo in Fife. It was Scotland's first nautical golden age.

By the late fifteenth century, Leith was the Queen of the Scottish ports with Andrew Wood's *Flower* riding proudly among the shipping that clustered along the Shore where the Water of Leith meets the Forth. The Kirkgate was the principal thoroughfare, a street thronged with merchants and skippers, wrights and labourers, sailmakers, artisans and craftsmen. Wood based himself in this centre of Scottish nautical activity, in a house nearly opposite that of the Barton clan. Not a seagull's cry away was St Mary's Church, recently built and significantly containing an altar to St Barbara, the patron saint of gunners.

After their disputed raid of 1481, the following year saw another English fleet venturing to the Forth, with Lord Howard and Sir Thomas Fulford in command; this time the Scots were waiting. Great sails billowing, the carvels would push out from Leith, and there would be gunfire and choking smoke in the Forth, harsh cries of fighting men and the flash of sun on Leith axes. With naval gunnery in its infancy, warships still preferred to grapple and board, so the multi-pronged hooks would fasten onto taffrail and bulwark as war trumpets rallied the battling sailors. It is unfortunate that there are no details of the encounter but the raiders were sent home with some loss, and Andrew Wood was, at least partially, responsible.

Acknowledging the part that Wood had played, in March 1483 King James III granted him a feu charter of the lands and village of Lower Largo in Eastern Fife. There was more reason than mere generosity in this grant, for the king had stationed his best seaman at the entrance to his main trading area, with a clear outlook on the shipping and a vested interest in preventing piracy. Yet for all his new prestige, Wood had not been born to command; he had gone to sea as an apprentice, clambering through the hawsehole to rise to become master and owner of *Flower*. Soon Andrew Wood would also charter the royal vessel, *Yellow Carvel*, so recently captured by the English. Knighted by the king and perhaps still in his mid-twenties, Andrew Wood skippered *Flower* on voyages to the Low Countries, Scotland's longstanding trading partner. His reputation as a seaman continued to grow, and when rebellion flared a few years later, Wood was to prove one of the king's most loyal servants.

Although history has affixed a turbulent reputation on Scotland, the country was not unduly prone to major civil war. However, on this occasion, the lands north of Forth mainly supported the king while the central Lowlands and Borders opposed him. The rebels used the teenage Prince James as a figurehead to justify their attack on the Crown with Andrew Wood's ships between the two forces, controlling the Firth of Forth. When the rebellion ended in the king's defeat and murder at Sauchieburn, Wood was powerless to intervene as his ships could not alter the course of a land battle. Tradition claims, however, that the seamen of *Flower* took some wounded royalists aboard. Tradition also claims that after the battle Wood was taken to see the victorious, and very youthful, Prince James.

'Sir; have you my father?' The young prince is reported to have asked.

'Sir, I have not your father' Sir Andrew replied 'but would to God he were on my ship, I would keep him skaithless.' (Unharmed)

Perhaps the prince was impressed. Certainly, he lost no time in befriending the loyal mariner.

Quick to take advantage of the perceived weakness of a nation with a minor on the throne, an English squadron of five ships struck out for the Forth. It seemed that the bad old days had returned when defenceless merchant shipping ran before the concentrated firepower of the pirates. The experienced English raiders took a heavy toll until Sir Andrew Wood with *Flower* and *Yellow Carvel* sailed to counter them. Tradition states that the rival squadrons met off Dunbar, where today the gaunt red shell of the castle boasts a colony of squawking kittiwakes and the nearby nuclear power station sits in surprising harmony with the coast.

With the odds so unequal Wood had to fight hard, and it is possible to picture the battle scenes. Although Wood's vessels carried cannon, the balls were too light to inflict serious damage, so while swivel guns fired onto enemy decks, the crews gathered to close and board. There would be bloody action with grappling hooks, pikes and swords, crewmen roaring murder beneath the swaying masts while the Saltire and the Cross of St George hung limply in the yellow smoke. Again the vicious axe of Leith with its long staff and wicked hook would engage in bloody slaughter. Perhaps time has sepia-tinted these old battles, but the fighting was in deadly earnest, the pain of wounds as acute, and the blood as raw and red, as in any similar terrorist attack of today.

On land, crowds lined the coast to cheer what they could see of the battle. Blood sports were the norm at that period, and what higher sport than man against man with death or captivity to the loser and a ship-full of goods for the victor? Eventually, the Scots gained the upper hand, and brought all five English pirate ships into Leith, to the delight of the persevering folk of that sunny port.

The English, however, are renowned for their refusal to admit defeat and Henry VII allegedly offered huge rewards to anybody who could restore what he probably thought was the proper order of things. According to the story, Stephen Bull came forward to put the Scots back in their place. Bull was a Londoner, reputedly one of the most notable captains of his time and he handpicked the crews for his three powerful ships. He had crossbowmen for the fighting tops, pike men to combat the formidable Leith axes and mariners experienced in vicious affrays in the Narrow Seas. He also knew he would face Andrew Wood. The English captain was no pirate out for easy prey but a hard fighting seaman hoping to defeat an enemy of his king.

Arriving off the Forth when Andrew Wood was trading in Flanders, Bull took up position in the lee of the Isle of May, captured a local fishing boat and forced the crew aloft to look out for his enemy. When two ships rounded St Abbs Head, the fishermen confirmed that they were Wood's *Yellow Carvel* and *Flower*, no doubt adding their own opinion of the outcome of the impending conflict. It was then that Bull did something that may already have been a nautical tradition. He called together his officers and drank a toast. The Scottish historian Pitscottie, who wrote colourfully and perhaps not always accurately, put this action more picturesquely, stating that Bull 'gart peirse the wyne and drank a toast with all his skippers and captains…' before clearing his ships for action.

Wood might have been surprised when an Englishman ambushed him in Scottish waters, but as a fighting seaman, he was probably used to such encounters. No doubt he had the cannon quickly manned, but crossbows, fireballs and pots of blinding quicklime were more suited to fifteenth- century sea warfare. With these weapons in the fighting tops and boarding pikes and two handed swords waiting on deck, Wood's crew prepared for action. The men would be tense as the rival ships closed, weather battered sails pushing them over the sea, gulls screaming astern and the wind a constant howl through the rigging. Perhaps somebody prayed, fingering his rosary beads, maybe a man cursed; an officer would snap orders to bring his vessel closer to the wind while the savage cliffs of May Island and the hills of Fife loomed on the starboard bow.

At this period seamen had not yet experienced the joys of West Indian rum so that wine was the standard nautical tipple. In an action strikingly and perhaps suspiciously similar to his opponent, Sir Andrew Wood 'caussit to fill the wyne and everie man drank to wther' (caused to fill the wine and every man drank to another). Then the crews ran to their stations, ready to face this enemy who was so much like themselves. As the sun rose, the rival squadrons squared up, weather-battered ships rising and swaying on the blue chops of the firth. The Scots had the advantage of the easterly wind, but their cannons were outranged so when the more powerful English guns opened fire they could not retaliate. With shot screaming over his ship and raising tall columns of water nearby, Wood put on all sail to close with the enemy. His men would have to endure a length of time under fire before they could utilise the hand-held weapons they knew so well.

Two ships against three, all manned with veterans who were eager to fight while overhead flapped the serene blue and white Saltire and the bold red cross of St George. Grappling hooks flickered, thumped onto English oak, onto Scottish decks and yelling seamen clashed with sword and boarding pike. Crossbow bolts hummed, quicklime showered agonisingly, but the English were picked fighting men, and they far outnumbered the Scots. The battle continued all that day, ceased out of exhaustion at night and restarted in the pale of dawn as trumpets blared to rally the battered crews.

With the ships locked together by grappling hooks and the fighting so intense that the seamen failed to notice spectators crowding the coast, the ships drifted north into the Firth of Tay where the sandbanks proved a natural defence. The English vessels, with their deeper draught, ran aground and were unable to manoeuvre as Wood's crewmen attacked. Stuck in the sand as the tide ebbed, the English surrendered. The Scots towed all three of their vessels into Dundee, and Sir Andrew Wood handed Bull to James IV, who chivalrously sent him home to England. After all, there was official peace between the two realms.

Next year, 1491, the king allowed Wood to build a castle on his lands at Largo, and from here he kept watch over the Forth. Although Stephen Bull had been speedily repatriated, his crewmen were less fortunate. Wood employed them in hacking out a canal that stretched from the Kirkton of Largo to his castle and only when the work was complete were the men returned home. It is still possible to trace the indentation on the ground.

That sea battle was perhaps the high point of Wood's career, but he was to sail again on the king's business, as well as on his own. In 1495 he was in Hebridean waters with *Flower* as King James took a fleet to daunt the proud chiefs of the west. Nine years later he was back when the king bombarded the fortress of Cairn Na Burgh in the Treshnish islands during the rebellion of Donald Dubh and visited his surprisingly loyal subject MacIain of Ardnamurchan. But in 1506 it was William Brownhill who accompanied the Earl of Huntly to the western seas, and the Barton family did not recruit Wood in their private feud with the maritime might of Portugal. Nor did Wood take part in the campaign in the Baltic to aid King Hans of Denmark, while it was Andrew Barton who cleared the North Sea of Dutch pirates.

Despite this apparent neglect, when the king built the great *Michael* at Newhaven, Andrew Wood was in overall charge of the construction. *Michael* was a huge ship with a hull strong enough to withstand shot from heavy cannon. With Wood in command, great things might have been achieved, but when she sailed from the Forth, Wood was not even aboard. Social status, it seemed, had superseded both sense and experience. However Sir Andrew Wood lived to a well-earned retirement. On his devout journeys from his castle at Largo to the local church, eight members of his old crew rowed his barge along the canal his prisoners had dug. Rather than the perverse waves of the North Sea, it was the fertile countryside of Fife that witnessed Wood's last voyages, but even in old age, Andrew Wood had both character and style.

The greatest ship afloat

Although many Scottish kings were more of a curse than a blessing to the nation their presence supposedly graced, there was the odd monarch who shone like a diamond in a Lothian coalfield. Malcolm II was one such, a warrior king who defeated both English and Norseman to secure his kingdom. Robert I was another, and James IV could have been a third, although Clan Donald may disagree.

King James IV had an inauspicious start to his reign when he backed a rebellion against his father and a bitter end when he led a powerful army to foolish defeat at Flodden. In between he raised Scotland to a nation far more important than her size should allow, enhanced cultural activities and created a tradition of shipbuilding that would come to real fruition more than three centuries in the future.

King James IV was a maritime minded king. In June 1492 and in March 1504 his parliaments passed Acts that advocated coastal burghs to build twenty-ton fishing vessels. These boats were to carry 'all nettin and uther necessar graith convenient for the taking of greit fische and small' (all netting and other necessary gear convenient for the taking of great fish and small). Nor did the king neglect the crews, for 'stark iddle men' were to 'pass with the said schippe for thair wages.' Although the principal reason may have been to create employment for these idle men, James may well have had an ulterior motive, for fishing bred a reserve of seamen that the king could tap in times of war and James was as much a captain as a king.

In this as in so much, King James showed foresight, for more than two centuries later, Arctic whaling was also partly conceived to build up a pool of experienced seamen.

From this small beginning, the king moved on to shipbuilding. James intended his first newly-built vessel to be used in the west, for in 1493 the Mac-Donald Lordship of the Isles had been forfeited, much to the disgust of the Hebridean. That same year, as the king sailed to the Isles, Tarbert Castle was repaired and a new royal castle built at Campbeltown. Six Leith shipbuilders travelled west to supervise local men building a new galley at Dumbarton, a port that also repaired the royal ship, *Christopher*. However, neither Dumbarton nor Leith had the facilities to build the great ship that James hoped to build. For a ship of the size James envisaged, the shipwrights would need deeper water. Casting around for a suitable spot, James paid royal gold to the monks of Holyrood Abbey for land west of Leith, and perhaps in the monk's honour named the area 'Our Lady's Port of Grace'. James had a pier and dockyard built, erected a small village and filled it with imported workmen from half the nations of maritime Europe. While the educated referred to this busy, graceful port as *Novus Portus de Leith*, the locals, more prosaic called it the Newhaven. Blunt speaking was always the way around the Forth, so Newhaven it has remained.

What grew at Newhaven was a phenomenon, a masterpiece of the shipbuilders' art. She was named *Michael*, after the patron saint of seamen, but she has gone down in history as *Great Michael*, arguably the most famous ship to come out of early modern Scotland and a foretaste of the wonders to come. She was at least four years in the building, traditionally used all the woods and forests of Fife except the royal hunting preserve of Falkland and when finished must have dwarfed the cluster of huts and houses that grew in her shadow. One thousand tons weight and about one hundred and sixty feet from stem to stern, her ten- foot thick hull was impervious to cannon fire; legend says that James proved this by firing a cannon at her, to see the iron shot rebound harmlessly. Formidable in defence, she appeared less impressive in attack with only sixteen small cannon on each broadside, two in the stern and one pointing forward. Her three hundred strong crew also had personal armament; culverins, double dogs and crossbows. There were four gunners for each of her deck guns plus a complement of soldiers; fully manned, *Michael* would be an extremely crowded ship.

It was totally in character for King James to haunt Newhaven. He joined the shipwrights at their work, listened to a storyteller, befriended the ubiquitous children and joined the local fishermen when they rowed to the oyster beds of the Forth. On the day of *Michael's* launch, James had supper on board, with lamplight reflecting on the Forth and the music of a Highland harper accompanying the sound of seabirds and the hush of the waves. Sea trials proved that *Michael* was cumbersome, and her deep draught meant she was apt to run aground, a constant danger at a time when men had not yet created accurate charts.

But she was a symbol of a proud nation and floated off Inchkeith with the royal standard fluttering proud and shields of the nobility lining her gunwales. Perhaps too, she was a symbol of the chivalric dreams of the young king, for James spoke of sailing her to lead a crusade against the Turks in the Holy Land. There is something sad in the romantic hopes of James, proud king of a small country, who lived with the dreams of a past age while thrusting forward into the Renaissance. It is quintessentially Scottish, this blending of past and future, this King who grasped at something magnificent while mingling happily with the lowest of his people, but the romantic dream of *Michael* was most unlike the pragmatic character of the Forth.

It is perhaps not surprising that *Michael* was not to achieve the deeds the king had hoped of her. Launched in 1511, she only had two years to grace the Forth before the nearly inevitable English war began. It was not unexpected, and the fault did not lie entirely with the impetuous James. France became embroiled with an English war and the romantic, brave and honourable Scottish king could not do less than support his ally. On land the unruly Borderers were repulsed from Berwick; at sea, the Frenchman De la Motte captured seven English ships: the war had begun.

Leith was Scotland's maritime centre and the home of the fighting Barton clan. It was only a couple of years since Andrew Barton had waged a successful war with Portugal, the nautical super power of the time, but now he was dead, killed by English cannon after he practised piracy on a nation who were acknowledged experts at the trade. Others of his family put to sea. John Barton sailed from France with war supplies while Robert Barton brought captured English ships into Leith, where they sulked beside the Shore in sullen defeat. Robert was himself no maritime novice, having already served as a Scottish naval commander in the expedition to aid King Hans of Denmark against Sweden.

Meanwhile, the French proposed that they should borrow the painstakingly built Scottish navy for fifty thousand francs. Not surprisingly, *Michael* was their particular target. James, however, was not inclined to give away his ships. Instead, he whistled up the Earl of Arran and placed him in command of the fleet. It was customary to have a high-ranking noble in command of naval and military expeditions, and Arran was an experienced commander who had led the fleet to the Hebrides in 1504. James accompanied them as far as the Isle of May before sending them out of the Forth. Not southward, to support the army that was advancing across Tweed, but north about, heading for Ireland and possibly France. Scottish history might have been very different had James sailed with his fleet rather than leading his army.

Although it seems strange to reach France by sailing in the opposite direction, there was logic in this movement. While the Scottish fleet was perhaps eleven ships strong, the English had twenty-four vessels, manned by mariners every bit as experienced and warlike as the Scots. If Arran had tried to reach France by the direct route, he would have to fight all the way. Instead, the fleet rounded Scotland, threaded through the Pentland Firth and down the Hebridean Sea to attack Carrickfergus in Antrim. Rather than a piratical stroke for plunder, this was a blow at England's domination of Ireland and another example of James' loyalty to his allies; his fleet was supporting Hugh Roe O'Donnell of Tyrconnel who had also requested Scottish help.

With the smoking wreck of Carrickfergus left astern Arran led the fleet up the Clyde to land the loot. After some time spent among the fleshpots of Ayr, the fleet set sail again, to encounter autumnal gales that tested the seamanship of the crews and the seaworthiness of the ships. *Michael*, deeper of draught than the others, ran aground.

King Louis had expected the Scottish fleet to arrive in late August. The lone escort he had arranged, *Petit Louis*, endured a fortnight of storms before she returned to port. The Scots finally sailed into a French port at the end of September. By then the war had been lost. Romantic and far too impetuous, King James lay dead amidst the bloody shambles of Flodden. Already the French and English were negotiating peace as English raiding parties turned the Scottish Borders into a smoking charnel house.

It was November before the fleet returned to the Forth, but they left *Michael* and two other vessels behind. In April 1514 an Act of the Lords of Council reported her as being sold for 40,000 francs, and after that, she seems to have disappeared from history. Legend has her engaged in sea battles, or sinking as a burning wreck, or simply rotting away in Brest. *Michael* never fulfilled her potential, but in her way, she was a harbinger of the future. In time Scotland would produce many of the finest ships in the world, and certainly, *Michael* had been great as she swung at her anchor in the lee of Inchkeith with heraldic shields lining her gunwales and the rampant lion roaring from aloft. But other great ships were not so welcome in the Forth.

The Spanish are coming

It must the most famous failed invasion in western history, and perhaps with due cause. After all, the antagonists were extremely unequal. On one side was Spain, the most powerful country in Europe, a nation that controlled much of the continent and the bulk of the Americas, a nation whose galleons dominated the known seas of the world. On the other side was England, a little state that occupied slightly over half an off-shore island, with no colonies and a chequered history of aggression against her even smaller neighbours. Spain was Catholic, England was Protestant, and for years their vessels had clashed at sea.

So Spain launched her famed Armada to crush the English. They named it the *felicissima* or *invencible*, most fortunate or invincible, and the Duke of Medina Sidonia led it out there were around one hundred and thirty ships dedicated to defeating England. The Renaissance world had never witnessed such a display of maritime power; the sight alone should have been enough to cow any smaller nation.

With mariners such as Raleigh and Drake under the command of Lord Howard of Effingham, the English did not flinch; they met the invincible Armada at the extreme range of their cannon and harried the crescent-shaped Spanish fleet along the Channel. Gunsmoke drifted across the treacherous seas, and cannon fire echoed from the white cliffs of the Channel. When the Spanish anchored off Calais to pick up soldiers, the English sent in fire ships, and in the ensuing panic, the rival fleets clashed off Gravelines. Beset by foul weather, harassed by their elusive opponents, the Spanish headed north.

With so many ships on the loose, the newly Protestant people of Scotland watched the sea with some alarm. Already, in May, King James and over two hundred of the nobility had decided to repel any attempted Spanish landing; fighting ships were waiting at Leith and Dundee, while the authorities ordered Jesuits and other Roman Catholics out of the kingdom. On the 8th August 1588, when the Armada was in the Channel the Lord Admiral of Scotland, the Earl of Bothell, put to sea, and by the 10th August, English ships reinforced the Scottish vessels that waited in the Forth to ambush any Spaniards that strayed from the fleet. When a solitary Spanish ship sailed in, she was promptly snapped up.

James Melville, the minister of Anstruther, wrote about the worries of coastal Scotland. 'Terrible was the feir' he said, and described some of the rumours where 'sum tymes' the Spanish landed at Dunbar 'sum tymes at St Andrews and in Tay, and now and then at Aberdein and Cromertie Firth.' Melville and his contemporaries in Anstruther were well aware of the danger that could strike from the sea. Only two years earlier men from the East Neuk had chased English pirates as far south as Suffolk, while with any ship a possible predator, every voyage could be hazardous. In this case, however, the Spanish had no intention of raiding coastal villages, let alone invading Scotland. Reeling from their reception in the Channel, and buffeted by foul weather, the Armada staggered up the Scottish coast.

Their plan was to sail around the north of Scotland and return to Spain, but their bad luck continued. The weather did not improve. Ship after ship was forced ashore, to die on the wicked rocks of western Ireland, and crew after crew perished in the howling surf or under the merciless knives of the locals. Others sunk off Scotland. In October forty-six shipwrecked Spaniards, survivors of *La Ballanzara*, staggered into Edinburgh, to be fed and comforted and returned home in *Grace of God* and Andrew Lamb's *Mary Grace*. But Anstruther remained undisturbed by such events and the good people of the burgh must have relaxed a little as the autumn months passed into winter and the chances of a Spanish landing lessened.

So it must have come as something of a shock when a frantic man awakened Melville one dark December morning. 'I have to tell you news sir!' Melville would have clambered groggily out of bed as he tried to clear his head. The man continued, no doubt gabbling out his tale. There was a ship full of Spaniards in the harbour of Anstruther. It had sailed in that morning, but there was nothing to fear, for they were seeking help, not intending to burn and loot and slaughter. Well, that was bad news tempered with good, but Melville knew that he could make no decision on his own and he gathered all the 'honest men' of the town and arranged a meeting with the leader of the Spaniards.

It is not hard to imagine the tension as the Kirk minister advanced toward the representative of Catholic Spain. There was a brisk December wind raising spindrift from the chopped waves of the Forth, and smoke from the chimneys of Anstruther would waft across the waterfront. The Spaniard too would surely be nervous, aware that he was isolated in a strange land, responsible for the lives of his men. He would wonder what sort of reception he would get from this grave faced man in this cold Scottish town. Scotland was Protestant, Spain Catholic, but there was no history of hostility between the two nations. On the contrary, there were some Catholic Scots including Captain Sempill, who had attempted to persuade Spain to use Scotland as a base from which to attack England. After all, Scots and English had been enemies for centuries, and it was only a few years since England's Elizabeth had had Mary Queen of Scots judicially murdered. Indeed, earlier that year a Captain Keble of Harwich had been fighting Scottish seamen at Havre, and the Scots had promised to throw him or any other 'English dogs' overboard if they caught them at sea. So it was not all bad news for the Spanish.

Melville thought the Spaniard a 'verie reverend man of big stature' as he opened the conversation with a torrent of Spanish. Most of the Anstruther men would listen in bafflement, but Melville was educated and understood the visitor's language. The Spaniard related a fascinating tale of woe.

He was Captain Juan Gomez de Medina of *El Gran Grifon*; a 650-ton ship built in Rostock. *El Gran Grifon* was not a galleon, but the flagship of a squadron of transports, and the majority of her crew were German although she also carried Spanish soldiers. Once the English had hounded the Armada out of the Channel, *El Gran Grifon* sailed north, sometimes with other vessels in view, often seemingly alone on the churning North Sea waves. By the end of September they had passed the east coast of Scotland and rather than sail through the Pentland Firth with its ferocious tides, Gomez sailed further north, to attempt the channel known as 'the Hole' between Fair Isle and Shetland. It was a dangerous procedure, sailing west against the prevailing winds in waters that would be familiar to very few. The ship was leaking badly and out of food when a look-out sighted land.

'At last' an unknown diarist recorded 'when we thought all hope was gone except through God and His Holy Mother, who never fail those who call upon them, we sighted an island ahead of us.' It was Fair Isle, half way between Shetland and Orkney, bleak, barren and battered by the weather, but *El Gran Grifon* was in no condition to continue. Gomez manoeuvred as close as he could to the rocks of Stromshellier, at the south east of the island. The upper yards of the vessel were scraping the top of the cliffs of Fair Isle. It was a feat of superb seamanship as the captain sacrificed his ship for the sake of the crew.

As the ship ground against the rocks and the great grey seas pounded her timbers, two hundred crewmen and perhaps a hundred soldiers clambered aloft and inched along the yardarm to crawl to safety. Assembled on dry land, they marched westward until they reached a settlement where the locals looked at them in awe. Not used to men in gaudy clothing, yet alone to men in glittering armour, the islanders from all seventeen inhabited crofts believed the Heavenly Host had visited them. The unknown diarist thought the islanders were dirty and savage, but they were also hospitable and gave everything they could to these strange men who needed help. Unfortunately, the island could barely raise enough food for themselves yet alone for such a massive incursion, and both Spaniards and Fair Islanders struggled with something approaching starvation. Perhaps fifty of the visitors died, probably as a result of the exertions of the voyage plus famine and disease. It was seven weeks before Andrew Umphrey of Berry brought a boat, or perhaps boats, to carry the rest of the survivors off Fair Isle, and it was early December before the weather battered Spaniards arrived at Anstruther.

Now Gomez looked hopefully at Melville and said they wanted to 'kiss the King's Majestie's hands of Scotland.' After enduring so much, the Spaniards hoped for help in the cold Protestant land. It was that very Protestantism that acted as a barrier to any great friendship, as Melville explained, but next day the Spaniards were allowed ashore. If the people of Anstruther held any lingering vestige of fear, they lost it when the 'young beardless men, sillie, trauchled and houngered' (Silly, bedraggled and hungry) arrived, to wander bewildered among the grey houses that were so different from those of their homeland.

Eager for news of the rest of the fleet, Gomez was visibly upset when he heard about the wrecks that littered the coasts of Scotland and Ireland, but he was relieved to be safe in Anstruther, as would his crew. After a while, they dispersed, with the noblemen invited to the court of King James, who ordered the Provost of Edinburgh to look after the ordinary mariners. Scottish people helped some, but hunger forced the remainder to beg in the grimy Edinburgh streets. Others remained in Anstruther, where they were cared for as best the small settlement could. After all, Anstruther was a maritime town, and these were shipwrecked mariners, despite any doctrinal differences.

It was summer of 1589 before four ships left Leith with over six hundred Spaniards on board; another hundred left in November, after spending months fighting in Maclean of Duart's private wars.

On his return to Spain Gomez called in at Calais, to find some Scottish seamen languishing in the local prison. After making enquiries, he found they were from Anstruther, and he had them released, taking with them his compliments to the Laird of Anstruther, to his hosts and to the Protestant minister in that Fife town. Even in times of bitter religious disagreement, Gomez proved that he was a Spanish ship's captain and a true gentleman in the best sense of the word.

But perhaps that was not the last of the Spaniards in Fife, for local folklore long claimed that some young Spanish seamen were made so welcome in the Anstruther homes that many children were born nine months later. These children had a slightly darker complexion than was usual, and a distinctive Spanish cast to their faces. It would be pleasant to think that there was such tangible proof of the Spanish-Scottish alliance, but perhaps that is only hearsay.

Chapter Two

The Outer Islands

The islands of the Firth of Forth add character and colour to an area already notably full of both. For the sake of clarity, in this book, they are arranged into two geographical groups, inner and outer islands. All have their own story.

The Isle of May

The Isle of May stands guard over the northern entrance of the firth, and at around five miles from the coast of Fife is the furthest out of all the Forth islands. It is large by the standards of the Forth, about a mile long by a third of a mile wide, and has a maritime history quite distinct from any other. May has seen sea battles and Norse raids, Celtic monks and smugglers, Jacobites, fishermen and a lighthouse tragedy, yet it is now a peaceful place to visit, resplendent with bird and animal life.

When seen from Anstruther or Crail, May appears to be a rocky island of cliffs, yet nowhere does it rise more than one hundred and fifty feet above sea level. Perhaps because of its isolation, May looks larger than it is, but boat trips from the coast of Fife deposit a regular dose of summer visitors. In rough weather the trip can be an adventure for May is as much an island of the exposed North Sea as of the sheltered Forth, but the rewards can be great.

As always in this part of Scotland, the origins of the island's name are disputed. May might have been derived from the Gaelic *Eilian Mhaigh*, a plain, or from the Norse *Maeyar*, island of seagulls; both Gaels and Norse had associations with the island and both left their mark. There is the Gaelic term, Tarbert, for an isthmus, towards the south of May, as well as Ardchattan, Point of the Cat, and St Colme's Hole, possibly from St Columba, although there is no record of that saint ever coming to the Forth. The term Kirkhaven may be from the Norse Havn or harbour, but may also be a purely Scots word with the same meaning. Either derivation augments the religious significance of this island, as does the place name Altarstanes. May was a pilgrimage site and a place of burial from time immemorial. According to Lothian lore, Thenaw, the mother of Saint Kentigern, floated to May on a coracle. Legend claims that Thenaw was one of nine priestesses who presided over the ancient settlement at Traprain Law.

Today, however, the Scottish Natural Heritage owns and manages May Island, and does a splendid job of caring for visitors and resident wildlife. They took over from the Northern Lighthouse Board in 1989, and both have added to the attractions of this fascinating island.

Despite, or perhaps because of, its geographical remoteness, ninth century Christians built a church on May. The foundations were on a much earlier burial mound that indicated people living on the island from perhaps the Bronze Age. A fragment of pottery found on the island dates from around 2000 BC, which may suggest human presence at that date, while flint arrow heads suggest that early humans hunted here. It is sometimes hard to realise that people so long ago lived in these remote and now deserted locations, while we congregate in vast numbers into cities where there is little space and only minimal contact with nature. Our lives are certainly more technologically advanced, but we must have lost something from our purely abstract brushes with the natural realities of the world. It is interesting to visit May Island and to contemplate an ancient settlement here, where the power of wind and sea is everything and humanity merely part of a much richer and more diverse world.

According to some legends, the Isle of May was home to at least two Celtic saints. The first was Saint Ethernan, who was a Pict who sailed to Ireland, became a priest and returned to be a missionary in Scotland. He died around 669 AD and was buried on the Isle of May, which became a place of pilgrimage in his honour. He also has links with a standing stone in Kilrenny in Fife. The second saint is Saint Adrian of May, who lived peacefully on the island until the Norse murdered him in 875. The present chapel on May is dedicated to his memory.

The church on May has a long history, being one of the many improved by King David I in the early twelfth century. Nine priests lived permanently on May, and their principal function was to pray for the soul of the kings of Scots. Given the often-bloody history of Royal Scotland, these devout men must have had a lot of praying to do. However, they also had other jobs, including caring for the hordes of pilgrims that descended on the island; we cannot tell the numbers, but the ten-seater communal lavatory argues for quite a few. It may have been the English wars of the late thirteenth and early fourteenth century that finally persuaded the monks to leave such a vulnerable spot, for some time after St Andrews took control of the priory from its erstwhile owners of Reading Abbey, the monks of May relocated to the safer environment of Pittenweem in Fife.

In the sixteenth- century the religious and political landscape of Scotland was undergoing significant changes and May exchanged hands several times before John Cunningham of Barnes gained ownership and around 1636 founded what was arguably Scotland's first lighthouse. Powered by coal, the light helped guide shipping away from the rocks of May, and the owner gained by charging a fee based on the ship's tonnage. While Scottish shipping paid two shillings Scots a ton, the rate was double for foreign vessels, but later lowered by around a quarter. This early lighthouse was relatively simple, a coal-powered beacon in a large basket. It was said to consume 400 tons of coal a year, which suggests that coal vessels must have been fairly regular visitors to the island.

Unfortunately, while helping save the lives of some, the coal light beacon created dangers for others. In January 1791, seven people were found dead within the lighthouse accommodation, one of the keepers, along with his wife and five of his children. The remaining child, a three-year-old girl, survived. An enquiry discovered that the keepers had been piling up the ash and clinker from the beacon until it reached the window of the keeper's room, and when a coal falling from the beacon sent the pile alight, smoke and fumes had suffocated the people inside.

Although the idea of a light beacon was excellent, its location and operation was not ideal. On 19 December 1810 two Royal naval warships, HMS *Pallas* and HMS *Nymphe* believed that they saw the beacon and steered accordingly, only for both ships to pile up on the East Lothian coast near Dunbar. They had mistaken the flame of a lime kiln for the May Island light. It was evident that a better system of lighting was necessary if May Island was not to endanger shipping. Accordingly, in 1814 the Northern Lighthouse Board bought the island for £60,000 and took over the operation of what was then the last private lighthouse in Scotland. Two years later Robert Stevenson built a more modern lighthouse, beautifully decorated with a gothic tower and castellated trimmings. The lighthouse had living space for three families, with extra accommodation for any officials who chose to visit, and in 1816 it began to operate.

Since that date, there have been constant improvements to the May Island lighthouse. The Northern Lighthouse Board added a new light and refractor lens in 1836, and there were further enhancements in 1885, with better accommodation and a boiler house. There was also an engine house with the then largest steam powered generators in Britain, at over four tons weight each. Despite a massive increase in power, the new light used only 150 tons of coal a year, less than the original open basket. In 1924 incandescent mantle oil light was fitted.

From 1843 May Island also had a secondary Low Light to keep vessels clear of the North Carr Rock, seven miles to the north of the May. This light operated until 1887 when the board set the North Carr Lightship in position. The third North Carr Lightship is presently in Victoria Harbour in Dundee.

As well as operating the light, the keepers were quite capable of more active rescues. The Firth of Forth was a busy fishing area, with boats from many of the Forth ports either chasing the herring or long-lining. In 1930 the trawler George encountered difficulties near the May, and two keepers jumped into the sea and swam to the rescue of the crew. Unfortunately, such help is no longer available since the light became fully automated in 1989, not long before the Northern Lighthouse Board relinquished control of the island to the then Nature Conservancy Council. Scottish National Heritage now owns the island.

Added to Celtic monks, Norse warriors and lighthouse men, the Isle of May has been home, temporarily or permanently, to other people. There was a village on the island in the eighteenth century, and it seems that the resident fishermen also indulged in smuggling, which was more of a national occupation than a marginal method of lawbreaking.

Perhaps the fishermen became involved in 1715 when a host of Jacobites descended on the island. That episode occurred during the Earl of Mar's Jacobite Rising when a small force under Mackintosh of Borlum intended a diversionary raid across the Forth and into Edinburgh, that several boatloads were stranded on May. They remained there for a while, quarrelling, before continuing their journey to Lothian.

Apart from that, May was often just off the mainstream of national politics. Three English vessels sheltered in the lee of the island before ambushing the Scottish mariner Andrew Wood around 1490, but Wood was the better fighter and captured them instead. The island was also close to the scene of the so-called Battle of May Island on 31 January 1918 when experimental steam powered submarines collided disastrously with other Royal Naval vessels, resulting in two sunken submarines and tragic loss of life. During the Second World War, the Royal Navy was back, with a control centre on the island for asdic to search for German submarines and surface vessels.

These events, however, seem to have been peripheral to the island. After the wars, it returned to its habitual windy isolation, a home for seals and, in high season, over 20,000 seabirds including the spectacular puffins, eider ducks, terns and guillemots. In season, fortunate visitors may stay at the observatory to experience the sights and sounds of this island, so near to Edinburgh in miles, yet so far in peace. Overall, the Isle of May remains a place for contemplation, and the early Celtic monks would approve.

The Bass Rock

For those visitors fortunate enough to arrive from the south, the distinctive shape of the Bass Rock is one of the most dramatic introductions possible to the Firth of Forth. Standing proudly a bare mile north of the fertile plain of East Lothian, the cliffs of the Bass owe much to the screaming colonies of gannets for their distinctive white colouring. The Bass is three hundred and thirteen feet high, with cliffs on three sides while on the fourth it boasts a surface that slopes to a landing place augmented by a castle and lighthouse. Throughout its eventful history, the Bass has been a fortification, royal refuge, royal prison and now a bird colony and Site of Special Scientific Interest.

Today most people will view the Bass as a dramatic piece of seascape, an island that adds to the view from the East Lothian Coast, or as a bird sanctuary. It is a place for visitors to admire, a rock for residents to remember while abroad and a treasured island for thousands of ornithologists. Indeed, there is nowhere quite like the Bass; in many ways, it is the most spectacular of all the islands of the Forth.

This section will begin with the gannets, for they have played, and continue to play, an important part of the story of the Bass. At present, there are around 150,000 gannets on the Bass Rock, around 60% of Europe's total and the largest colony of northern gannets in the world. They are incredible birds with an impressive wingspan of nearly six feet. More statistics only add to the gannet's reputation. They can dive at over ninety miles an hour – and to witness a gannet plummet into the Forth is a sight worth seeing. They also travel up to thirty feet below the surface of the sea and have been known to hunt over a hundred miles away from their home, returning across the North Sea to the Bass.

The formal name for the gannet is *Sulana Bassana*, after the rock on which they are so prolific, but in old Scotland, people knew them as Solan Geese. Despite the nomenclatural association, the bird is only a visitor. Gannets remain on the Bass between January and October, flying to the Gulf of Guinea or perhaps the Mediterranean for the dark days of Scotland's autumn and early winter. Although ornithologists now enjoy the antics of the gannets, they were once more valued for their food value. From the middle ages until the nineteenth-century gannets were harvested, with their eggs and flesh regarded as valuable food products. The nuns of North Berwick reaped their tithes from each barrel of gannet fat, and in the eighteenth-century gannets cost four shillings each in Edinburgh's Flesh Market. In the nineteenth century crowds of spectators watched as men bludgeoned the birds to death on the high cliffs, and their bodies plummeted to the sea below, where watchful boatmen collected the harvest. Even Queen Victoria, it was said, enjoyed her gannets' eggs from the Bass.

Today the Seabird Centre at nearby North Berwick has cameras on the island, and there are boat trips from the port that circle the island and even land there if the unreliable Forth weather allows such extravagances.

While the birds have presumably always been there, human activity stretches back to at least the eighth century AD. The first known inhabitant was a Celtic holy man known as Saint Baldred, one of many semi-forgotten Celtic saints. Also known as Balthere of Tyninghame or sometimes as the Apostle of the Lothians, Baldred operated primarily in East Lothian man, with churches stretching from North Berwick to East Linton and possibly a monastery at Tyninghame. In common with many Celtic missionaries, he was probably Irish born before moving to Holy Island in Northumberland, the Celtic centre for spreading Christianity to Northumbria and neighbouring states.

When not spreading the Word in the mainland, Baldred sought solace and the spirit of the Lord on the Bass Rock, although there are other sites on the East Lothian coast that bear his name. Celtic monks were famous for finding isolated islands on which to live, with May and Inchcolm also sharing the Celtic connection. Baldred's cell in the Bass is long gone, with the present ruined church being no older than 1493 although it may have been built on the site of his cell. At that time James IV, a notably Christian king, was on the throne.

There is another ruined St Baldred's church near Tyninghame House, which is traditionally the site of his monastery. The Norse, experts in early mediaeval terrorism, apparently sacked the monastery in 941; their ugly reputation for selecting soft targets to ravage seems well deserved.

Perhaps there were more visitors to the Bass after the death of Baldred, but it was because of the Norman infiltration into Scotland that the rock again became prominent. Queen Margaret, the Saxon wife of Malcolm III, and her son King David were instrumental in inviting many of these incomers into the country, often to marry some indigenous heiress. Unlike England, which the Normans conquered within a few years after their dramatic arrival, their influence in Scotland was more gradual and more piecemeal. However, they made a significant impact on the nation and Scotto-Norman names are part of mainstream Scottish history. Bruce, Stewart and Fraser are three that immediately spring to mind.

According to tradition, Malcolm III handed the Bass over to the Norman family of Lauder in the eleventh century, although there is another story that it was not until the late thirteenth century that the Lauders obtained the Bass. This alternative legend says that Sir Robert de Lawerdre had fought the English off the Tay and in return Wallace granted the family the Bass Rock. William Wallace, it must be remembered, had close links with Tayside, being supposedly educated in Dundee. Lauder appears to have been an ardent patriot who also fought alongside Wallace at Stirling Bridge. While the English judicially murdered Wallace in 1305, Sir Robert survived a further six years. He is buried in the Auld Kirk graveyard in North Berwick, within sight of the Bass.

Whether it was Wallace or King Malcolm who granted the Bass to the Lauders, it seems that they held the rock in return for the high rent of a white wax candle, which was to be presented to Tyninghame Church every Whitsunday. Even more unusually for Scottish nobility, many of whom would sell their granny for a shilling, the Lauders seem to have been a trustworthy bunch. The Lauders of the Bass retained their title and lands throughout the turbulent middle ages, which says much for their staying power if nothing else.

At the beginning of the fifteenth century, Scotland had a minor as heir to a vacant throne, which was only an invitation for the nobility to squabble and the neighbours to peer through the letterbox to see what they could steal. In 1406 the regent sent the very young King James I to the Bass, not as an exile, but for his safety. The idea was for James to wait for a ship to transport him to France, where he would remain until he came of age to be king. Sir Robert Lauder of the Bass was reputed to be a trustworthy man who would not betray his monarch, and so it proved, for Sir Robert looked after James and saw him safely onto a ship.

Unfortunately, there were others who played another hand, including the Duke of Albany, who alerted the English. Despite it being a time of peace, an English vessel pirated James's ship and transported him to London, where he was held hostage for the next nineteen years. James was a full grown man when he was eventually permitted to return to Scotland, but he remained a friend of Sir Robert Lauder.

King James found his kingdom a violent place, with various nobles causing turmoil, and he vowed to create peace. Remembering the sanctuary of the Bass, he sent some of his subjects there, either as prisoners or as royal hostages for the good behaviour of others. Perhaps the most significant was Neil Wasse Mackay, the teenage son of Mackay of Strathnaver. Today not many people visit that broad strath in the far North West of Scotland, and, since the notorious clearances of the Duchess of Sutherland, even fewer live there, but in the fifteenth century the Mackays could field thousands of fighting men. Even more importantly, young Neil Mackay was also related to MacDonald, Lord of the Isles, a Hebridean potentate so powerful he could rival the king. The Bass was holding a potentially very valuable prisoner.

For all his youth, Neil Wasse Mackay had already been involved in the battle of Harpsdale, one of a series of bloody encounters between the Mackays and Gunns that turned northern Scotland into a nightmare for any ordinary peace-loving person who lived there. For nine years, from 1428 until 1437 Neil Mackay, known later as Neil of the Bass, lingered in his island prison, but when his nobles murdered King James in Perth, Neil Mackay either escaped or was released. He made his way to Strathnaver to claim the chieftainship of the Mackays. The contrast between the confines of a tiny island and the wide sweeping strath could hardly have been greater.

His time on the Bass had not quietened Neil Wasse Mackay and as soon as he arrived in his homeland, he was deeply immersed in clan politics. He was present at the skirmish at Dounreay and the Battle of Sandside, where the Mackays slaughtered the Gunns. It seems that the king was justified in keeping him a prisoner on the Bass for so long.

For a century and more the Bass remained unsullied save for the annual harvest of the gannets and the constant crash of waves, but in the middle of the sixteenth century, an English king once more attempted to interfere in Scottish affairs. King Henry VIII of England had decided that his son should marry young princess Mary of Scots. When the Scots refused his kind offer, he sent his armies north to persuade Scotland that there were benefits in a Tudor marriage. His methods of wooing the young bride included orders that the Scottish Borders were to be 'tourmented and occupied as much as they can be'. Also, his soldiers were to 'raze and deface' Edinburgh and 'sack Leith and burn and subvert it and all the rest putting man, woman and child to fire and sword'. In fact, all the loving embraces of a potential father-in-law or an English monarch.

Henry's armies did their best as they murdered, looted and burned their way through the south of Scotland in this 'rough wooing', but when they came to the Bass, they stopped. The garrison of a hundred men, formidably installed within the castle, beat off two of their attacks and glowered at them across the mile-wide moat. Other Forth islands, less well prepared, fell to English arms.

The Lauders remained in favour at court, with George Lauder a Privy Councillor of King James VI, who was so impressed with the Bass that he offered to purchase it. However, the rock remained in Lauder hands until well into the seventeenth century. It was debt, not war that saw the Bass leave Lauder ownership, but its primary functions remained unchanged. Relatively close to Edinburgh and the centre of Scottish affairs, it was also secure and isolated, making it a perfect place for a castle and prison.

By the later seventeenth century the Stuart kings owned the rock and used it as a prison for the notoriously intractable Covenanters. These devout men and women refused to allow any monarch to dictate their religion, clinging to Presbyterianism where God was head of the Church rather than acceding to the Episcopal faith with a secular king in command. The Duke of Rothesay, Lord Chancellor of Scotland, responded with persecution and intolerance, sending dragoons to enforce the royal religion and deepening the devotion of the Covenanters by imposing hefty fines on some and condemning others to the gallows. Around forty were sent to the Bass, to learn respect for the crown by suffering confinement within the dank dungeons of the ancient castle.

The dungeons of the Bass were not pleasant, and the jailers did their best to make it as uncomfortable as possible. The governor charged his guests for their keep, so they paid for the privilege of imprisonment. If they had no money, they were fed dried salt fish and denied the fresh water that the guards commanded; instead, they drank stagnant water from puddles in the rock, filtering the foulness through oats to remove some of the impurities. Even the guards were hungry in inclement weather when the supply boats could not battle through wild seas. They took out their frustrations on the most intractable prisoners, lowering them into the deepest dungeons to suffer intense cold and constant dampness.

The names of the prisoners are a roll call of Covenanter honour. Among the best known were Alexander Peden, the prophet who roamed the south west during the worst of the Killing Times and John Blackadder, minister of Troqueer, who died on the Bass in 1687. Not far behind were Sir Hugh Campbell of Cessnock and Gilbert Rule who later became Principal of the University of Edinburgh as well as Alexander Gordon of Earlston. The list is daunting if testimony to the courage of these martyrs to the Presbyterian cause.

It was Peden who wrote most memorably about his experiences on the Bass:

We are close shut up in our chambers, not permitted to converse, diet, worship together, but conducted out by two at once in the day to breathe in the open air. Envying with reverence the birds their freedom, provoking and calling on us to bless him for the most common mercies, and again close shut up day and night to hear only the sighs and groans of our fellow prisoners.

But even the worst of times come to an end, and when William of Orange replaced James VII as king, the persecuted became the persecutors and men of the Presbyterian religion controlled the nation. It was now the turn of the Stuarts to be underdogs, and from cruel oppressors, history made them into romantic victims carrying the label of Jacobites.

The Jacobite war of 1689 saw the battles of Killiecrankie, Dunkeld and Cromdale when the Highlanders fought their rearguard action for a dynasty that had never shown them any favours, but people often forget that the last Jacobite stronghold in Scotland was the castle of the Bass.

With the king's cause crumbling all around, the Bass held out for King James. The same situation that gave them strength was also their weakness, however, for if an island fortress is hard to assault, it is also difficult to supply, and in 1690 the garrison was starved into submission. From being a Jacobite stronghold, the Bass became a Williamite prison, and staunch Jacobites were crammed into the cells that had so lately held Covenanters. Whatever accusations can be levelled at the Jacobites, lack of spirit is not one, and when the Governor, Fletcher of

Saltoun, was absent from the island, the four captives waited until their guards were unloading coal, and then they seized the castle.

Reinforced by men from the mainland, supplied by France and augmented by raids on the shores of East Lothian, the Jacobites held the Bass for four desperate years while the Williamites scratched their heads in despair and worried about attacks on their ships in the Forth. There were various incidents to which history has added a gloss of colour, but which must have been a bloody reality to the participants.

For example, there was the case of the Jacobite supporter named Trotter who the Williamites brought to the mainland shore opposite the Bass to be hanged. When the Williamites erected the gallows, the men of the Bass sighted their cannon and fired, scattering the execution party and temporarily saving Trotter's life. The Williamites executed him later, a martyr to the cause of a king who probably was not even aware of his existence.

By 1694 King William was becoming frustrated with the continuing Jacobite presence so close to his northern capital. He sent the entire strength of the Scottish Royal Navy – two frigates – to reduce the fortress, but their cannon proved powerless against the stone walls of the Bass. Once again it was starvation that forced surrender, but the defenders had a final trick to play. Inviting the Williamites on to the Bass, they produced a feast fit for a king, with the last of their food and wine, an appearance of plenty augmented by a troupe of dummies to give the impression that the garrison was powerfully numerous.

Either fooled or just fed up with the pinprick presence of the Jacobites, the Williamites allowed the Jacobites to leave unmolested and with full military honours. The sixteen Jacobites withdrew to France, possibly to join in the plots against the new regime, and the Bass finally slipped away from the centre of the national stage.

The history continued, of course, as the Crown relinquished control to private ownership and the castle gradually crumbled. David Stevenson planted a lighthouse instead, using the worked masonry of the castle for a better building, so now a beam of safety gleamed across the mouth of the Firth of Forth. Another and even more famous Stevenson also knew the Bass, for Robert Louis Stevenson holidayed in North Berwick as a child and placed his hero, Robert Balfour on the rock in one memorable scene in his book Catriona.

Today, however, nobody lives on the island. The Bass is left to the birds, those enduring, spectacular gannets that swoop and dive and survive on the white cliffs so close to the shores of mainland Scotland.

Craigleith

Only a few hundred yards from the town of North Berwick, Craigleith is possibly the first island that many young holidaymakers ever see. It is small, little more than a grassy rock, but still interesting to view, and, like many islands of the Forth, it is a well-known place for birdlife. Craigleith is one of a four-strong group of four islands: the others are the Bass, Lamb and Fidra.

Puffins, those sea parrots with the colourful beaks, breed on Craigleith, and at one time there were over 28,000 pairs, but the intrusive tree mallow has drifted across to the island and curtailed the colony by choking the burrows. After a terrible dip in puffin numbers, work by teams of volunteers to remove the mallow has seen an increase in numbers.

There was a colony of rabbits on Craigleith, bred specifically for food, and there are still seals that visit, to sun themselves in safety, so close to the bustling town of North Berwick. Apart from that, the island has a quiet history, perhaps because it is too small to have seen any of the colourful or bloody events that make the Forth islands so fascinating. However, it remains an excellent piece of seascape to admire from the beaches of North Berwick, and there is nothing wrong with that.

The Lamb

This tiny islet was always uninhabited, for there is little here and very little reason even to land, yet alone stay. Lamb is a small hump of rock, a plug of some ancient volcano, and with a tidal swell of some fifteen feet on an islet with no jetty or any other facility, landing is not easy. Lamb lies between the much more interesting islands of Fidra and Craigleith and has two outlying islets of North and South Dog Islands which are just rock skerries. Yet it does add to the scenic splendour of the seascape, and for that reason, it should be included in any book about the Forth.

Fidra

Fortunate is the person who first sees Fidra while walking from Dirleton to the beach at Yellowcraigs. If he or she comes on a bright day in spring or early summer, with a clear sky and the birds calling, then his or her fortune is doubled, for

Fidra is arguably the most scenically favoured of any island in the Forth. It has everything that a small island should; mysterious rocks, a lighthouse, a story, birds and seals, and beaches on which it is possible to land a small boat. Indeed, Fidra is that most enviable of all things, an accessible island with a feeling of remoteness and splendid views. If there was ever a heaven in Scotland, then here it sits.

The name could be Gaelic, from *Fiodra*, or Norse, *Fiorey*, but in this case, it does not matter. Fidra is Fidra, and whether the names means Feathery Island from the number of birds, or something completely different, it is still worth looking at, visiting, speaking about or just being.

At present Fidra is uninhabited and is used as a reserve by the Royal Society for the Protection of Birds, but at one time this was a working island. There is a jetty on the west side of the island, which the monks who manned St Nicholas's ancient lazaretto presumably used. It is pleasant to think that this beautiful place had such a useful past, for a lazaretto was a hospital, a place of goodness and caring.

Today there is an automated lighthouse and cameras that send images of the bird life to the seabird centre at North Berwick. However, it has a far larger claim to fame than many islands ten times its size, for there is a belief that Robert Louis Stevenson used Fidra as a model for Treasure Island.

The idea is attractive, and plausible, for Stevenson knew this area well. One can imagine RLS walking the golden sands of Yellowcraigs dreaming up his most famous story while watching the arches and cliffs of Fidra and listening to the pounding of the Forth surf. All he would need was a parrot, squawking that eternal 'pieces of eight,' and a one legged man with a smile to charm the wallet from your pocket.

Although small, Fidra is enchanting. Once seen, it will live in the memory and in the heart, immortalised not only because of Stevenson's eternal words.

Eyebroughty

This tiny islet is also known as Ibris and sits a mere couple of hundred yards from the coast of East Lothian. It is possibly the least considered of any of the islands of the Forth, but its situation alone makes it worth including. Belonging to the parish of Dirleton, it is next to the golden beach of Gullane and is also a notable bird reserve frequented by cormorants. In common with every other island in the Forth, it has a unique beauty.

Chapter Three

The Emigrant Fleets

On the 17th July 1698, the Shore of Leith shuddered with multitudes of people. They peered out to the Forth, young and old, Kirk Elders and romance-dreaming youths, bitter-eyed soldiers and pompous merchants. Their sole topic of conversation was the five vessels that tossed in Leith Roads, waiting for the crew to raise the anchors so that they could sail to high adventure and, more importantly, high profits.

The ships belonged to the Company of Scotland, which had been launched to trade in the Far East. Originally founded in London by a combination of Scots and English, political and commercial pressure applied by the not-quite-so Honourable East India Company had first broken and then forced that dream to northern resurrection and relocation in Edinburgh. William Paterson, an idealistic financier with a unique vision, was the mainspring of this second company. As the founder, Paterson was a hero to the Scots of his day, but in the end, it was his dream that wrecked the company, that nearly bankrupted the nation and caused the death of many hundreds of innocent people. He dreamed of a seaport where he could exchange European goods for Asiatic, where he could barter the products of the Occident for the wealth of the Orient. After reading the journal of William Dampier, a Caribbean buccaneer and maritime adventurer, Paterson decided that Darien was the ideal place for this entrepot.

To any casual observer, the idea had merit. Darien was a narrow isthmus in Central America washed on the east by the Caribbean that opened the sea door to Europe and on the west by the Pacific that led to Asia. A port here could have easy access to both seas, and so could cut travelling time. It would also be accessible to the raw colonies on the eastern seaboard of North America. But deeper delving would have revealed another side to this trading paradise.

Darien was on the Spanish Main, and that fact was surely enough to raise a concerned eyebrow. The Spanish had claimed this coast for centuries; they would not be pleased with any foreign intrusion. Besides that, if the site was so favourable why had it not been settled already? Any local could have answered. Darien was a jungle controlled by a tropical climate and infested with disease, where mosquitoes clouded above the rotting vegetation, and vampire bats hung from the trees. The heat was oppressive and the humidity worse; altogether not the best place for inexperienced Scotsmen from the Northern Hemisphere.

But Paterson dangled the lure of Darien and the Company bit. The Old World was forgotten in favour of the New. There were sealed orders for the five ships rolling in the Forth, for none of the hopeful traders and settlers officially knew where they were bound. East Indies or West, it mattered little to them: the old Scots love of adventure for its own sake had called most together, mixed with a dusting of avarice.

Of the twelve hundred volunteers, three hundred were from the flower of Scotland's nobility and others included veteran soldiers, a pilot, translators who were to be paid around £4 a month, three ministers with an annual stipend of £120 and a printer. Most of the others hoped to settle on a promised fifty acres of Darien soil. A very few wives came too; God-fearing women venturing to one of the least Godly parts of the world.

But if the Company apparently chose the settlers with care, they were less particular with the cargo. The ships carried candlesticks and locks, bolts of thick serge and seventy-nine dozen worsted stockings. There were five hundred pairs of slippers and nearly fifteen hundred Scots bonnets, thousand upon thousand clay pipes and more than a thousand Bibles to balance the kid gloves and hunting horns, four thousand periwigs and fifty fish giggs – the list seems a hopeless confusion of paraphernalia. Yet there was more method than is immediately apparent. These goods were not to be traded to the Darien Indians, or to the Chinese, but to the European settlers of North America.

Paterson had been waiting for ten years to create reality from his dreams. His initial idea had been presented to the Hanseatic trading towns of north Germany, as he considered Scotland was too poor for such a risk. In this he was correct; the £400,000 which the Scots scraped together was about half the nation's capital and left little in reserve in case of emergency. Little money, little talent, less experience and virtually no backing, as Paterson's proposed colony had stirred the jealousy of the English and raised the hackles of King William, whom Scotland shared with the English.

The five ships were to sail into a hostile sea, land the colonists on a savage shore and build a wooden town in a sodden jungle. Even so, the crowds cheered madly that July day as the ships, *St Andrew, Unicorn, Caledonia, Endeavour* and *Dolphin* creaked away from Leith Roads, bound north-about for the unknown. The hopes of Scotland sailed with them.

At Madeira, the shipmasters opened their sealed orders, raised their eyebrows in feigned surprise and set course for the west, for this proposed Eldorado, but before they reached Darien disease had already claimed forty-four of the settlers. When they sailed into the long bay that was their destination, there were tall trees and thick bush, fresh streams and a hill to use as a lookout. And waiting to meet them were the insects; always there were mosquitoes, buzzing and biting the sweating flesh of the Scots. In the manner of myriad colonists of the future, they applied familiar names to their new home. The bay became Caledonia, the fort they built, St Andrews, and their shack village was New Edinburgh. A minor disaster struck as *Unicorn* ran aground on a rock, but they befriended the local natives, gave them guid Scots names, and for a while, things seemed idyllic.

Appearances, however, were deceptive. Landsmen quarrelled with seamen while malaria, yellow fever and the bloody flux killed both without discrimination; they did not complete the fort, and Caledonia Bay proved a trap. As none of the Scottish ships could sail against the prevailing winds, the weather hemmed them against a lee shore. Merchant ships entered the bay from the North American colonies, but the traders did not want any of the curious collection of goods the Scots had brought and sailed away. Seventy-six colonists died by Christmas, most of them poor planters who had come in the hope of land and found only a sordid grave. The colony began to starve. When *Dolphin* struggled out of the bay to find provisions, foul weather drove her ashore, and the crew disappeared into Spanish jails. And then the Spanish attacked the colony. But that might have been something of a relief to the Scots, for if they were helpless against disease and naïve at intercontinental trade, fighting was something they understood. Captain Thomas Montgomerie led a hundred men - probably all that remained fit in the colony - and thrust the Spaniards back through the steaming green jungle.

But a single jungle skirmish could only lift the colonists' spirits for a short time. Prodded by English traders who feared for the profits of their Caribbean colonies, and the Spanish who he valued as allies in his war with the French, King William disowned the Scottish settlers. Dispirited by frustration, decimated by disease, the colonists crept into the four remaining ships and headed home. It was a terrible voyage; *Caledonia* lost 105 dead, *St Andrew* 130 and *Endeavour* sunk.

Ignorant of the disaster a second expedition was sent out from Scotland. Fresh settlers rebuilt the huts of New Edinburgh and erected a perimeter fence, but the Spanish again advanced against the colony and this time they were in earnest. Campbell of Fonab rose from the tragedy of Darien and led a counter attack that threw the Spaniards back to the jungle. This battle of Toubacanti was the last military victory of an independent Scotland, but there were always more Spanish than Scots, they were organised and used to the climate. A siege began.

Fonab refused to parley, but others did not, and after weeks of despairing misery the colony collapsed. Perhaps if Fonab had been in sole command things might have been different, but it was not to be; the Scottish dream of Empire ended in a Caribbean jungle. There were very few of the settlers that saw Scotland again; Campbell of Fonab was one. Others ended as bonded servants on English Caribbean colonies, or died at sea, or settled in the Carolinas, or just disappeared. Paterson tried to keep his dream alive, urging Scotland toward the Union of 1707 but it was not until the second decade of the twentieth century that the Panama Canal proved that his idea was sound.

However, the demise of Scottish Darien did not mark the end of the Scottish connection with Central America. One hundred and twenty-four years after the first Darien fleet sailed optimistically from Leith, another band of emigrants slipped out of the same port to enjoy the delights of the tropics. If anything, their hopes were even more ill-founded, for while Paterson was at least genuine in his entrepreneurialism, the later settlers were the victims of a sardonic confidence trick.

The early nineteenth century was a time of utter confusion in South and Central America as three centuries of Spanish rule crumbled before a combination of local liberation forces and European mercenaries. There were Scotsmen involved, of course, for with the end of hostilities in Europe in 1815 thousands of unemployed soldiers searched for the only work they knew. Some had crossed the Atlantic even while Bonaparte's war continued, and one of these was young Gregor MacGregor. He was a twenty-year-old soldier of the Black Watch when he heard about the South Americans' struggle for liberation from Spain and, like so many Scots before and since he sailed away to fight somebody else's war.

Simon Bolivar was the Venezuelan equivalent of William Wallace, and he welcomed young Gregor, as he had welcomed Colonel Campbell and Mackintosh the saddler and all the other footloose Scotsmen into his army. But Gregor was distinctive; he seemed to have inherited the talent for unconventional warfare enjoyed by his outlaw ancestors and quickly rose to command first a brigade, then a division. Gregor was so successful and so personally dashing that he married Bolivar's niece, Donna Josepha. Even as a married man he kept fighting, at sea as well as on land.

Gregor's raids on the old Spanish Main smacked of Henry Morgan, or perhaps the pirate MacNeils, and it was at the height of his fame that he landed on the well-named Mosquito Coast. Britain had claimed this area from time to time, and log cutters had attempted to settle here, but the climate and the insects had defeated them, as they had defeated the Darien colonists. General of Division Gregor MacGregor, however, had no intention of settling in person. Obtaining a grant of the land from the indigenous chief, he sailed back to Britain.

Gregor boldly approached the new king, George IV, and announced himself as His Serene Highness Gregor I, Prince of Poyais, while his companion, plain William Richardson was presented as 'Commander of the Most Illustrious Order of the Green Cross, Major in Our Regiment of Horse Guards.' After a decade of daring deeds in the Americas, the self-appointed Prince of Poyais powered ahead. He issued currency adorned with the Clan Gregor motto, and he produced brochures that announced the wonders of his new kingdom. Rob Roy would have nodded his approval as Gregor tapped a London bank for £200,000 based on the non-existent gold and diamonds of Poyais. And then the Prince and Princess travelled north to Edinburgh.

Here he sold Poyais land and exchanged Poyais currency for Scottish banknotes while arranging for colonists to cross the Atlantic to his golden land. The first ship left in early 1822, the second a few weeks later and the emigrants found themselves struggling on the insect infested coast of Central America, like their predecessors over a century earlier.

A London newspaper published the following account, supposed to be from one of the settlers:

We landed on the 22nd April at the mouth of the Black River, walked about a furlong on its banks ... It was now past sunset; the fire flies had illuminated the trees, the grasshoppers and a numerous band of musicians of the feathered tribe had begun their nocturnal melodies; the myrtle and thickets of strange trees which I cannot describe overhung its banks, perfuming the air with their fragrance ... The climate is very fine; the mornings and evenings most pleasant. Potatoes and all the English seeds which have been sown thrive well; we had a good pleasant voyage; all of us landed in as good health as that in which we left Scotland. This is the most beautiful country I have ever seen.

The reality was different, as other papers mentioned that: 'individuals whom Sir Gregor Macgregor sent as settlers to his viceroyship of Poyais had arrived in that bay in a state of wretched destitution.'

Gregor had promised a civilised capital city and a country where they could luck gold for pleasure. Instead, they found untamed jungle and swamp. Seven more shiploads were on their way when the British Colonial Agent of British Honduras rescued the first group. The Agent, John Young, also sent a description of the actual nature of the Mosquito Coast. Gregor treated Young's attack with some scorn, but withdrew to France, to play the same confidence trick again. Not surprisingly, he was eventually cast into a French jail, from which he extricated himself and retired to a warm welcome in Venezuela. Perhaps it is national embarrassment that kept this launching of a second Darien scheme out of most history books, but for the duped settlers involved, it must have been a terrible experience. Other mariners from the Forth had different adventures.

Chapter Four

The Real Robinson Crusoe

In the late seventeenth-century ship masters from the Forth had to possess the Dean of Guild's Ticket and pay the guild's dues before they were allowed to leave harbour, and Scottish ships sailed to all the navigable coasts of Europe as well as the New World. They were men of stature, these shipmasters, respected in the communities of which they were pillars, men responsible for the lives of their crews and the safety of their cargoes and ships. Yet at sea most commanded vessels not much larger than today's inshore fishing boats, and on land, their homes were often so close to the sea that spray could rattle the windows.

These maritime towns of Fife were stone built, with solid, unpretentious buildings, often tarred as protection against sometimes atrocious weather. The pantiled roofs are nearly as common in the East Neuk now as they were then, but what is not so usual is the niche aboard a doorway in the village of Lower Largo, and the statue that was placed here in 1885. Many visitors stare at this figure, and not all know who he might be. It is an impressive carving, the figure of a man dressed in tattered goatskins, with a pistol thrust through his belt and a long musket in his left hand. His right hand is sheltering his eyes, as if from a tropic sun, and he peers anxiously out to sea. The stance is fitting for this is the statue, and the cottage replaces the original home, of Robinson Crusoe. Of course, there was no Robinson Crusoe, Mariner, of York, but there certainly was an Alexander Selkirk, Master Mariner, of Largo.

Alexander Selkirk, or Selcraig, must be unique in that an author and spy preserved his memory by writing a book and did not mention him once. And that raises an intriguing question – how many other mariners like Alexander Selkirk have there been, equally deserving of memory but who did not happen to meet a novelist of the calibre of Defoe? Perhaps the story of Selkirk will suffice for all.

Born in 1676, Selkirk was the son of a Kirk elder and shoemaker, but young Alexander had the wrong temperament to follow in his father's leather footsteps. As a youth he mistakenly drank some sea water and, when his brother laughed, as brothers tend to do, Selkirk thumped him. Such minor disputes happen in most families, but when Selkirk decided to escalate the fisticuffs into something more serious and looked for his pistol, his father moved to stop him. A violent man, Selkirk overpowered both his father and his elder, married sister. It was then that his sister-in-law intervened, with typical Fife diplomacy.

'Ye false loun' she is said to have yelled, 'will ye murder your father and my husband both?'

There was no murder, but the matter seeped outside the Selkirk family. The Kirk ruled supreme in old Scotland, and Alexander Selkirk was summoned to stand on the Stool of Repentance to be lectured at by the minister while the entire congregation watched, listened and no-doubt sniggered. After this ritual humiliation, running away to sea seemed an excellent idea.

The hot tempered Selkirk made such a good mariner that by 1704 he had risen to Sailing Master of the English *Cinque Ports*, a sixty gun, forty-ton ship with over sixty of a crew. *Cinque Ports* seems to have been a licensed privateer, carrying Letters of Marque from the government. These documents allowed her to plunder the king's enemies for the pursuit of victory, glory and personal profit. Such privateers were common in the early eighteenth century, and every maritime nation of Europe used them as a cheap method of augmenting their navies. However, the early eighteenth century was also the period of the freebooter when a seaman could begin a voyage on a legitimate merchant ship and end as quartermaster of a pirate, without changing vessels. It was not unusual for a crew, already suffering from the habitual hardships of life at sea, to remove a tyrannical master and replace him with one of their own choice. It was equally possible for a ship's master to remove a recalcitrant member of the crew.

At this time few non–Spanish vessels sailed to the Pacific from Europe. Those that did were nearly invariably pirates or privateers. *Cinque Ports*, commanded by Captain Stradling, was in company with Captain Damper's *St George* on a cruise along the Pacific coast of South America. They hoped to capture one of the two Spanish treasure galleons that sailed annually from the Philippines with a combined cargo valued at over half a million pounds. Like most such attempts, this one failed. A change in captains did not improve the morale of the crews, and there was near mutiny at the Juan Fernandez Islands. A few months later, off the coast of Peru, the two captains disagreed, and the ships parted company.

These were dangerous waters for a single English ship, for since the first Iberian mariners had pushed across the Atlantic and threaded through the strait between Tierra del Fuego and the mainland, Spain had claimed the Pacific for herself. Ignoring the danger, *Cinque Ports* slipped down the coast of South America, becoming less and less seaworthy with every nautical mile. Eventually, in September Stradling put back into Mas a Tierra, the largest island of the Juan Fernandez group, this time to refit.

With voyages that could sometimes last for years, eighteenth-century seamen were accustomed to caring for their vessels without recourse to dockyards. *Cinque Ports* was beached and careened, with the hands scraping off the accumulated marine growth that would slow and eventually damage the timbers of the ship's hull. The planking was then re-caulked and possibly coated with tar to make the vessel waterproof. Odysseus would have approved of this ancient practice. When this procedure was complete, Selkirk was still dissatisfied with the seaworthiness of the ship. He stated that he would prefer to remain on the island rather than chance the Pacific in a vessel in such poor condition.

Stradling, who appears to have been an unpleasant man, promptly agreed and marooned Selkirk on the island, with only his sea chest to keep him company. Selkirk's quick temper had cost him friendship, and if he had hoped for support, he found none. Stradling ignored his appeal for a change of heart and refused to allow him back on board. Possibly in a state of shock, Selkirk watched the topsails of *Cinque Ports* slip beneath the horizon, leaving him to mourn his fate.

As it happened, no mourning was necessary. Selkirk's Forth-trained seamanship was superior to that of his captain. While Selkirk explored his new home, his opinion of *Cinque Ports* seaworthiness was justified. Forced to beach in the Mapella Islands for urgent repairs, the Spanish captured her crew and Stradling spent the next six years entombed in various dungeons. Nonetheless, Selkirk pined for company on Mas a Tierra, which was no desolate, tiny islet but a sizeable chunk of land of some thirty-six square miles, green, hilly and well wooded. Even with freedom to roam, Selkirk was still marooned, sentenced to solitary confinement for an unknown length of time. In the present age of instant communications, it is difficult to imagine the feeling of sheer aloneness, with no knowing when, if ever, a ship might call at the island. Selkirk hardly slept, ate sparsely and continually watched the sail-less horizon. There must have been mixed feelings here, for with this entire ocean – and all the vast adjacent landmass – claimed by Spain, any ship that did appear would in all probability have been Spanish. Selkirk would have exchanged his island prison for the far worse conditions of a Spanish dungeon. Fortunately, Selkirk missed that experience. Instead, he read his Bible – his father had been a Kirk elder after all – and grew accustomed to his own company.

As time passed Selkirk became adept at this pioneering life; he built himself two huts, one for living and one as a makeshift kitchen, and lined them with skins. He ate generously off wild goats, wild cabbage and the turnips and parsley planted by earlier visitors. There was a never ending supply of shellfish but, perhaps providentially, no Man Friday and no raiding cannibals. In reality, although he may not have realised it, Selkirk had a better quality and standard of living than the majority of people at home, or in most places in the world. Free of disease, free from crime, with sufficient food and shelter and an equable climate, Selkirk only lacked human company and given his quick temperament that may have been no bad thing.

Selkirk remained on Mas a Tierra for more than four years, becoming so fit that he could not only outrun the wild goats but also catch and eat them. He was also so handy he could manufacture knife blades from the hoops that had held barrels together. All the same, he was still eager to light the bonfire he had prepared in case a sail should puncture the monotonous horizon. So when two ships did appear toward nightfall on the last day of January 1709, Selkirk put flame to his heaped up pile of brushwood. He must have prayed that the visitors were friendly.

The faint northerly breeze would fan the flames, the smoke would drift sweetly southward, but Selkirk would have suffered agonies when both ships slid away, white sails vanishing into the dark, but he was seaman enough to realise that they could not approach against the wind. Dawn brought fresh hope as the ships crept closer, anchored offshore and sent a yawl to the island. Were they friendly or hostile? The tension would surely have been sickening as the small boat approached with its crew of six armed men. Only when they landed would Selkirk be sure of their nationality. While he had been on the island, amazing things had happened back home. Separated by centuries of animosity, Scotland and England had merged in a lopsided Union that created Great Britain. Commanded by Thomas Dover, the yawl's crew was British, or rather a mixture of British and just about every other nationality at sea, but they were friendly. By a coincidence that was not so amazing given the tight knit world of the South Sea privateer, Selkirk knew one of them.

Both ships, *Duke* and *Duchess*, were under the command of Captain Woodes Rogers of Bristol, but one of the officers was William Dampier, one-time captain of *St George*. Dampier was one of the most interesting characters of a fascinating age. He was an Englishman from the West Country, and he had also once been marooned, on the Nicobar Islands, while his chequered career had included spells as a logwood cutter, buccaneer and explorer. Although accused at one time or another of cruelty, drunkenness and cowardice, he always bounced back to prominence. Dampier seemed to have a knack of touching on Scottish maritime affairs, being loosely associated with the Scots colony at Darien, which in turn had a tenuous connection with Captain Kidd. Now he greeted Alexander Selkirk.

Hirsute, with quiet eyes and broken speech, Selkirk made an unusual but still welcome addition to Woodes Rogers' officers. At least he was fit and certainly still a seaman. But Selkirk was not destined to enjoy a quiet passage back home. Perhaps that was for the best, for his few years sojourn on a Pacific island had not tamed him.

Duke and *Duchess* were privateers, out to circumnavigate the globe, damage the Spanish and, if possible, capture the two elusive treasure ships that annually crossed the Pacific. When Rogers seized a fifty-ton prize, he named her *Increase* and appointed Selkirk master. Other prizes followed and then an expedition thirty miles up the Guayas River to Guayaquil. The privateers were after easy plunder, not hard fighting and when the British captured some Spanish ladies, searching them was not an unwelcome duty.

Some of their largest Gold Chains were conceal'd, and wound about their Middles, legs and Thighs etc, but the Gentlewomen of those hot Countries being very thin clad with Silk and fine Linnen... Our Men by pressing felt the Chains etc., with their Hand on the Outside of the Lady's Apparel, and by their Linguist modestly desired the Gentlewomen to take 'em off and surrender 'em.

Searching scantily clad women must have made a strange contrast to the monastic existence on Mas a Tierra, but somehow Selkirk survived.

There were no casualties on this jungle expedition, but fever followed them out to sea. The voyage deteriorated, with discontent among the hands, talk of a prisoner's uprising, the birth of a child to one female slave and the whipping of another for what was termed 'immodest behaviour.' Overall, these incidents were probably nothing out of the ordinary for a privateering cruise, and any discontent was forgotten when a vessel was sighted. She proved to be one of the hugely valuable Spanish treasure galleons on her annual passage from the Orient to the New World. With laughter and anticipation, the privateers readied their weapons.

Incarnacion was the smaller of the two Spanish galleons and Rogers captured her with just two casualties. A wood splinter pierced the buttock on one unfortunate seaman, but more importantly, a musket ball wounded Rogers in the jaw. An attack on the larger galleon failed spectacularly, but *Incarnacion* was rich enough for any but the greediest of privateers. Although unable to properly articulate, Rogers was still in charge and he chose Selkirk as one of her officers. So it was in a position of authority that the Largo seaman accompanied *Incarnacion* on her slow crawl across the Pacific. He would see the beauties of the Dutch dominated Spice Islands and would be party to the discussion that saw the ship join a Dutch convoy to Europe and, eventually, London.

Selkirk's world circumnavigation must be one of the most unusual on record, while Rogers account *A Cruising Voyage Round the World* allowed three pages to the Fife privateer. The English novelist and one-time spy, Daniel Defoe, read the pages and, according to legend, met the Largo man in a Bristol tavern to hear his story. The outcome was Robinson Crusoe.

Selkirk's share of the prize money was £800, and in 1712 he arrived back at Lower Largo, immaculately dressed in gold laced clothing and so changed that at first, nobody knew him. He married a local woman and tried to settle down, but this was not in his nature. Maybe it was his time on Mas a Tierra which drove him to seek out the more remote parts of Fife at the expense of time with his wife, but eventually, he had to leave home again. Some half-understood compulsion drove him to it, with his restless nature allowing him no peace. In a way his actions mirror those veterans of overseas war who return home only to seek solitude in the lonely places, hoping that external peace can restore tranquillity to a disturbed mind and soul.

Leaving behind his gun, his clothes and a coconut shell drinking cup that he had used on Mas a Tierra, Selkirk said goodbye to his mother and wife.

The ocean of tears he also left might have weighed on his conscience but if so he did not show it as he spent time and money with an English prostitute. That, however, was not the answer and, inevitably, Selkirk returned to his older love, the sea. Perhaps he could only find peace amidst the constant motion of a ship, or in the total solitude of his island. Whatever the reason, Selkirk became Lieutenant of HMS *Weymouth*, and in 1721 he died of fever off the West African coast.

Alexander Selkirk has not been forgotten. Augmenting Defoe's book is the poem by William Cowper whose lines include the well known 'I am monarch of all I survey' that is supposed to relate to Selkirk's island kingdom. If Selkirk looked out to sea from Lower Largo, he would have seen the islands of the Forth, and perhaps thought of the island he knew as home.

Chapter Five

Jacobites in the Forth

While Alexander Selkirk was indulging in foreign adventures, the people of the Forth found there was enough excitement on their doorstep. Firstly there was the Union with England, then the associated threat of Jacobism. The Union was not exactly the most popular political procedure in the world; indeed while there is evidence that influential people in England had bribed the Scottish lords into agreeing to the union, the people of Edinburgh, Glasgow and many other places greeted the news with angry riots. They felt betrayed and wanted their country back. In a time long before politicians considered democracy, the elite ignored the wishes of the majority.

Only a year after the Union, the Jacobites - supporters of the ejected Catholic Stuart monarchs - tried to exploit the situation. Naturally, they were less concerned with repealing the Union than with replacing a Hanoverian king with one from the Stuart dynasty, but their timing was right, for Scotland was seething with genuine anger. As so often in the complicated relationship between Scotland and England, France played Devil's Advocate, or faithful ally, depending on one's point of view. On this occasion, France supplied the Jacobites with six regiments of French infantry and one of Irish, thirty fast privateers and five men of war. It was a formidable force and was led by the redoubtable Admiral Forbin. Nobody doubted the nautical skill of Forbin; he was a privateer who had captured powerful men of war, a seaman who had sailed to the Arctic to attack Dutch and English fishing vessels and a warrior who had taken part in fleet actions. Legend claims that Forbin was the descendant of a Forbes, one of the many thousands of Scots that fought for France in the Middle Ages, and legend may be right, but Forbin and the fleet waited in Dunkirk and St Omer for the arrival of a king.

Invasion

For once, there was a king who wanted to oblige, but when James VIII, the so-called Old Pretender, arrived in March, he promptly caught measles, not so easily cured then as now. Bad luck usually sat on the shoulders of the Stuarts, but at least he had arrived, spotty or not, and Forbin could think about departure. His destination was Burntisland, where James Malcolm of Grange was to act as a liaison between the French and a Jacobite force. It was an Irishman by the name of Nathaniel Hooke who had smoothed the path of invasion. Carried in the French frigate *L'Audacieuse*, commanded by a Scottish captain named James Carron, Hooke had sailed to the Jacobite Countess of Erroll at Slains Castle, only for the British warship *Royal Mary* to intercept him. It seemed as if the Jacobite plans would be stillborn, as the French and British ships sailed broadside to broadside on the North Sea, the crews edgy, with men nervous at their guns and the seagulls circling overhead. Then the Countess appeared on deck and hailed *Royal Mary*. Captain Thomas Gordon came forward to listen to her.

'Could you kindly sheer off?' The Countess requested politely, as men spat on tar-stained hands and matches smouldered in their tubs. There could have been mayhem, but instead, Captain Gordon turned *Royal Mary* away.

Gordon was an interesting man. Possibly Aberdonian by birth, he had commanded merchant ships to the Baltic, Norway and the Netherlands. In 1703 he became Captain of *Royal Mary* in the Royal Scots Navy, sailing out of Leith to protect the east coast, a task at which he seemed to excel. In the summer of 1704, Gordon captured the French privateers *Fox* and *Marmedon* and brought them into Leith, with more prizes in the autumn and next year. Despite his illustrious nautical career, Gordon may have been a closet Jacobite or perhaps he was susceptible to a woman's voice, but either way his actions with the Countess allowed clear passage to James's invasion plans.

However, the best-laid plans of mice, men and kings aft gang agley, particularly when the royal Stuarts were concerned, and the next obstacle was already cruising in the Channel, hoping for a battle. It must have been very frustrating for the French and Jacobites, for between them they could devise some fine-sounding plans, with invasion and insurrection, armies landing and combining and the downfall of the Hanoverian dynasty virtually guaranteed. But then the Royal Navy stepped in and spoiled things. And here they were. Thirty-eight

British warships led by Admiral Leake, eager for a sea battle and blocking the Pretender's route to his destiny, or to Burntisland at least.

But for once the weather was on the Stuart side, producing a nasty little gale that blasted Leake back toward the Downs and providing Forbin with a window of opportunity to leave France. With James aboard the flagship *Le Mars*, the French were hopeful for reaching Scotland, until the continuing bad weather forced them to shelter at the Dunes of Ostend at the coast of the Spanish Netherlands. Gales damaged three vessels, so Forbin pressed on with a depleted fleet and a feeling of foreboding.

The storm continued. Forbin sent *La Protee* ahead to meet Malcolm of Grange and must have cursed as gales thrust the fleet northward, well past the Firth of Forth. Forbin struggled south with the wind howling in the rigging and most of the soldiers on board retching with seasickness. Even to an experienced mariner like Forbin, there must have been a sensation of relief when at last he entered the relatively sheltered mouth of the firth and anchored between the Isle of May and Crail. His relief would have turned to frustration when he learned that instead of Malcolm, *La Protee* had met a boatload of fishermen. There was no Jacobite army waiting to welcome the French. And worse, much worse, the Royal Navy was back, sailing astern of the French, blocking their route home.

Admiral George Byng commanded this second British fleet. He was a solid fighting seaman who had taken part in the capture of Gibraltar in 1704 and who had seen action at Malaga and Messina. He was no novice to be brushed aside. Byng ordered his ships into line astern, an invitation to battle that was hard to ignore. But Forbin knew that he commanded privateers and not professional warships; he was a commerce raider and guerrilla fighter by instinct, and was very aware that his ships could not face the Royal Navy in a gun to gun action.

When the enemy was still forming into their battle line, Forbin clapped on all sail and ran north, his swift privateers outpacing the warships as they slipped through the passage between May and Fife Ness. Or most of them did. The slowest French vessel was *Salisbury*, which, ironically the French had previously captured from the Royal Navy, and together with her consorts *Le Griffin* and *L'Auguste*, she fought a duel with the foremost of the pursuing British vessels. It was a stubborn fight, but not an altogether successful one for the following day, 14 March 1708; the fifty-gun HMS *Leopard* ran alongside and boarded *Salisbury*. In a situation like this, the Royal Navy was seldom defeated and soon controlled the French ship. It is another irony of history that the captain of *Leopard* was none other than Thomas Gordon, late of *Royal Mary*. A decade later Gordon refused to swear allegiance to King George I, resigned from the Royal Navy and joined the Russian service, where he rose to Admiral and became Governor of Kronstadt.

But the last words of that failed Jacobite attempt of 1708 undoubtedly belong to Private Deane of the Foot Guards, who was one of the unfortunate British infantry who was transported back and forth across the North Sea to counter the expected French invasion.

While we lay on board we had continual destruction in the foretop, the Pox above board, the plague between decks, hell in the forecastle, and the Devil at the helm.

There was no romance for the men involved in the Jacobite campaigns, but still, they tried.

Broadswords over the Forth

While popular historians and novelists have paid much more attention to the efforts of the Bonny Prince in 1745, the rising of 1715 was potentially far more dangerous. Led by Bobbing John, the politically unprincipled Earl of Mar, it commanded more support and occurred at a more sensitive time, than that of Prince Charles. With the Union only seven years old, the memory of independence was still fresh and the knowledge of new taxes very bitter. These new taxes were regarded, not without cause, as anti-Scottish. On the other hand, the Jacobites had reasons for optimism; the French had already nearly succeeded in invading Scotland and confirmed Unionists only narrowly defeated an attempt in Parliament to repeal the Union.

Many Scots considered the return of a Stuart monarch as the solution to their woes, while even in England there was some Jacobite support. The Earl of Mar, embittered by the lack of gratitude following his exertions in securing the Union, bobbed over the political fence in an attempt to recover the realm for the Stuarts. Nowhere was as disaffected as the Highlands, so Mar invited the known and suspected Jacobite chiefs to a deer hunt and sounded out their opinions. Very soon he raised the standard of rebellion at Braemar, very near to where the annual Highland Gathering is still held, attended by the present royals.

The initial success of the Jacobites was startling and must have created unease among the people beside the Forth. Cameron, MacLean, Drummond, the various septs of Clan Donald, Mackintosh, even the northern clan Mackenzie, declared for King James and when William Mackintosh of Borlum, 'Old Borlum', captured Inverness, it seemed the Jacobites could conquer all of Scotland. Mar himself marched on Perth, garrisoned the town and waited for reinforcements. However all was not as rosy as it looked; lowland Scotland had little love for the men from the glens, viewing them with a mixture of antipathy and fear. And with reason, for the Highlanders were ferocious warriors whose military prowess had frequently swept aside all opposition.

The Hanoverians also had an agenda. Clan Campbell was adept at backing a winner and its chief, the Duke of Argyll, known as 'Red John of the Battles', controlled the crossing of the Forth at Stirling, the strategic centre of Scotland. A veteran of the continental campaigns of Marlborough, Argyll commanded the government army in Scotland, three thousand redcoats and dragoons. Tough men, trained with musket and bayonet, the redcoats would recall previous engagements with the Highlanders and hope for revenge. If Mar wanted to move south to the Lowlands or England, he must first fight or outflank Argyll.

For all his political shifting, Mar was no fool, and he sent the Master of Sinclair into Fife to scour the East Neuk for boats, hoping to gather sufficient numbers to send a sizeable force across the Forth. Possibly the idea was to divert attention from the main Jacobite army or to make Argyll split his forces. Brigadier Mackintosh of Borlum would command this diversion. Borlum was a tall man of about sixty, stern faced, and grey eyed. He lacked the finesse thought necessary to act the courtier, as the court of the exiled King James in St. Germaine discovered when Old Borlum visited. People likened him to a bulldog for his manners, but this could be a backhanded compliment for a stubborn fighting soldier with a broad Scots accent.

Borlum commanded around two and a half thousand men when he left Mar's main army at Perth. Most were Highlanders; Drummonds and his own well-armed Mackintoshes, but there was also a regiment from Angus, men as willing to unsheathe a sword for King James as any red-shanked Gael from the glens. On the 11th October 1715, the Jacobites marched into Fife. No doubt they were watched by the Fifers, men glowering at them with mixed suspicion and fear, women either hiding, afraid of their reputation, or possibly hoping to catch the eye of a handsome Highlander. Either way, the Jacobites slipped through the byways poured into the small fishing villages of the Neuk, proclaimed King James VIII and readied themselves for the crossing.

The Fife fishermen were hardy, but they were men of the sea, accustomed to storms and salt wind; they were not fighting men fit to face Highland broadswords. So there was no resistance when the Jacobites arrived in a swirl of tartan philabegs and the clatter of steel. One detachment of five hundred flooded into Burntisland threw up barricades and loudly announced their imminent departure for the south shore of the Forth. Naturally, this activity attracted the cruising patrols of the Royal Navy warships, as was the intention, and three government sloops, backed by armed pinnaces crossed the Forth from Leith Roads to hover offshore. When the Burntisland Jacobites boarded their small craft in a noisy diversion, the Navy engaged them in a brief, pointless fire fight.

No doubt the folk of Burntisland watched as the powder smoke rolled in yellow-white clouds above the Forth, and hoped that both sides would go away.

Probably pleased at the success of his strategy, Borlum waited for darkness on the night of 12th October – perhaps around six o' clock – and embarked about a thousand of his men in commandeered East Neuk boats. It would be a tense time, with the Highlanders with their claymores and Lochaber axes brushing shoulders with the men of Angus as they filed into the boats. The night was cool, dark and probably windy and the Fifers would watch resentfully through half-shuttered windows or would shelter behind closed doors. Their boats were not just a possession; they represented the livelihood of the fishermen, the only means of feeding their families.

As the seagull flies, the distance from the East Neuk to the coast of Lothian is about eighteen miles. As the Jacobites had to row across, progress would be slow, with nervous glances to the west, from where the Navy might yet appear if they broke off from the Burntisland diversion. Royal Naval sloops would act like pikes among the minnows of the Jacobite fishing yawls. However, the Forth remained quiet, and after perhaps a few hours the keels kissed the welcoming sands of Lothian. From North Berwick to Aberlady the coast curves in quiet coves and wide bays; the Jacobites would clamber ashore to see the boats withdraw for the return journey. There would be soft splashes in the dark water, perhaps a muffled curse as a man missed a stroke and the gleam of moonlight on an incautious oar.

By daylight of 13th October the first Jacobite wave was well inland, and when an advance party probed into Haddington, an excited rider raced to inform the government forces. The Navy ceased its fruitless guard on Burntisland to sail eastward, toward the Jacobites. There would be cursing from the quarterdecks as the commanders realised that the enemy had fooled them and pigtailed petty officers shouting orders that squeezed every possible piece of speed from their craft. The Navy sloops arrived off the coast of East Lothian while the second wave of Jacobites was in mid crossing and a game of hide and seek began amid the skerries and offshore islets of the Forth. The Navy captured one boatload of forty men and drove others onto the rocky island of May, where the resident smuggling community may not have been too pleased to see them. The remainder of the Jacobites crossed safely, using the lights of the Navy's sloops as beacons to guide them safely ashore.

There were about three hundred Jacobites marooned on May, Highlanders and Lowlanders under Lord Strathmore and they fortified the island against the patrolling British warships as best they could. As they waited, tensions ran high, and the Highland Drummonds bickered with the men from Angus until the wind drove the Navy into the North Sea and the small boats cautiously probed out. But rather than head toward Borlum's main force in Lothian, they steered back toward Fife. The men of the East Neuk would have been relieved to see the return of their boats.

Borlum's force enjoyed mixed success. They found Edinburgh too well defended to attack, but captured Leith and freed the forty Jacobite prisoners. When government troops dithered, Borlum marched to Seton, leaving a trail of drunken men in his wake. From Seton, he headed south, joined the Earl of Derwentwater and penetrated as far as Preston. The Jacobites withstood a siege by redcoats but, on the orders of Derwentwater, surrendered. The results were predictable: brutality, imprisonment, banishment, slavery and execution. Borlum, however, was not a man to lock up with impunity. On the 4th May 1716, he led a mass break out from London's Newgate prison, charging down the guards before he entered the filthy warrens of London. Bulldog he had been termed, but it took more than dog-like tenacity to reach safety in France, it took cunning and skill. Borlum could do no more for the Jacobite cause, however, for by then the rising had terminated.

Other Jacobites, like minded and as ingenious, overpowered the guards on the ship taking them to slavery in the Americas and settled instead in France. Borlum's adventurous career continued when he joined the nearly-forgotten 1719 rising. His sons did not share his loyalty to a doomed cause. They emigrated to Georgia and joined the Highland Rangers there. But the people of Fife would be immensely pleased that the whole sordid business was over so they could return to their lives. In eighteenth-century Europe, however, periods of peace were merely preludes to the next war.

Chapter Six

The French Wars

When people think of the wars with France, they usually will think of Napoleon Bonaparte, the threat of French invasion on the English south coast or Admiral Nelson and Wellington. However all parts of Great Britain were involved in that sequence of wars, and the Forth was on the front line.

There was a guard ship stationed in Leith, and the Royal Navy was a constant presence in the Forth. The navy was busy, escorting convoys to the Baltic, searching the Arctic for privateers, patrolling the North Sea and bringing in prizes. Less romantically the Navy also landed prisoners-of-war to be transported inland to Penicuik and Edinburgh, and there was the hated Impress Service that scoured the towns on both sides of the Forth for reluctant recruits.

Father of the Russian Navy

While many people know that the Scottish born John Paul Jones is known as the Father of the United States Navy, fewer realise that a Scotsman also founded the Russian Navy.

In common with so many Scottish seamen, Sir Samuel Greig (1735-1788) was a Fife man, born in Inverkeithing and bred to the sea. With his father a ship owner, the young Greig went to sea in the coasting trade, learning his navigation and nautical skills in the tricky waters and sudden squalls of the North Sea. He left the merchant service and signed on as a Master's Mate in the Royal Navy. He first saw action at the capture of French island of Goree, off West Africa, and was in *Royal George*, then one of the largest ships in the world at the blockade of Brest and the Battle of Quiberon Bay in 1759. For a while, his captain was John Campbell who had been born at Kirkbean, not far from the birthplace of John Paul Jones. Further afield Greig was at the capture of Havana in Cuba, which was a major blow to the Spanish in the Caribbean.

With the end of the Seven Years War in 1763, the Royal Navy reduced its manpower and Greig was out of a job. Luckily Catherine the Great, Empress of Russia needed experienced officers, so Greig applied, and Catherine accepted him as a lieutenant. Promotion soon followed, first to captain and then to captain. By 1770 he was the commander of a squadron in the Mediterranean, facing the Ottoman Empire. Greig was heavily involved in the Russian victory over the Turks at the Battle of Cesme in 1770. There was one incident when Greig, alongside another Scotsman, Lieutenant Drysdale, led in a flotilla of fire ships against anchored Ottoman ships, set them ablaze and swam back to the Russian fleet.

Not surprisingly, Catherine promoted Greig to Rear Admiral. However great his achievements in war, it was probably Greig's peacetime accomplishments that were more significant. Catherine appointed Greig Grand Admiral and Governor of Kronstadt St Petersburg's naval base. He also became Knight of the Order of St Andrews, St George, St Vladimir and St Anne. Between 1774 and 1778 Greig reorganised, trained and disciplined the Russian Navy, pushing it up the ranks from a third rate service to a navy able to compete with the Swedes in northern waters.

Sweden was a powerful nation and Russia's rival for domination in the north. Greig led his newly-trained navy against the veteran Swedes near Hogeland in the Gulf of Finland. On a day of nerve-shredding gales, Russians and Swedes exchanged broadsides among the islands of the gulf until daylight faded. At the end of the conflict, the raw Russian seamen had acquitted themselves well, although Greig was not so happy with the quality of his officers. After a night of recovery, Greig sent the most inefficient officers ashore and sailed out again, seeking the Swedes.

With Greig in command, the Russians were in the ascendancy. He blockaded the Swedes in their base at Svendborg, the naval base of Helsinki and remained at sea. When he died, still on duty, he was awarded a state funeral. History remembers him as the father of the Russian Navy, and his son, Alexis created the Russian Black Sea fleet that faced the British during the Crimean War.

The Forth Smacks

Among the shipping that crowded the Forth were the Leith smacks. These were cutter rigged vessels with a large spread of canvas that sailed from Leith to Berwick and then to London with cargoes that included the Scottish staple export of salmon. By 1803 there were two smack shipping lines that operated the Leith to London route; the Leith and Berwick Shipping Company and the Edinburgh and Leith Shipping Company. Although they were merchant ships, the latter company had armed vessels, and the crews carried Protections, so the Royal Navy could not press them.

Sometimes the smacks had an interesting time, such as the voyage of *Britannia* and *Sprightly* in October 1804. They were sailing together for London when a French privateer attacked them off Cromer. The smacks kept close company and fired back, answering the Frenchman with shot for shot until she hauled her wind and ran away. Both smacks were damaged but quite able to sail, with not a single casualty.

According to the *Scots Magazine*, in January of the following year, a smack named *Swallow* was also attacked off Flamborough Head, which was a favourite hunting ground for privateers. The sea was busy then, with Geordie colliers carrying coals from Newcastle, but a fourteen-gun French privateer threaded through the shipping, closed with *Swallow* and opened fire from fifty feet.

But Leith smacks were never easy prey. Captain White ordered his men to man the carronades, which were short range, large calibre weapons made in Carron on the Forth, and nicknamed 'smashers'. They were carried on most British warships and used in the close quarter action so beloved by the Navy at the time. As the French vessel was close, the carronades were very efficient, and the passengers joined the crew in grabbing muskets and giving the privateer a hot reception.

The smacks continued their dangerous voyages until the end of the war, with two new companies formed. The voyages of these smacks were advertised in the press, for example in the *Caledonian Mercury* of 25 January 1804:

At Leith for London
The Union Shipping Company's Smacks
EDINBURGH PACKET
Wm Hall, Master
And

SPRIGHTLY PACKET
James Taylor, Master
Will take in goods, the former till this evening, and sail tomorrow after-noon, at two o' clock; and the latter till Thursday afternoon, at four o'clock, when she will sail.
Union Shipping Co's office
Leith, Jan 24, 1804

The Old Shipping Company advertised 'Armed Smacks' *Caledonia Packet* and *Queen Charlotte*, 'each armed with six carronade, eighteen pounders and four long four pounders.' The smacks remained armed until the end of the war. For example in April 1812 the *Caledonian Mercury* ran this advertisement:

Edinburgh and Leith Shipping Company's armed smack Forth, Thomas Tul-loch, master, now taking on goods, deliverable on Downe's Wharf, London, sails on Friday 3rd April at five in the afternoon.

Sometimes the smacks carried other things apart from mere 'goods'. In November 1813, when Britain was fighting America as well as Napoleonic France, some Dutchmen, ex-prisoners of war, had volunteered to join a newly raised battalion of the 60th Foot and fight their former allies rather than lan-guish in prison camps. They marched on board the smack Forth bound for London and glory.

Less martial men were also taken, such as the fifteen convicts from Glasgow who were chained and carried by coach to Edinburgh, to be shipped on Forth for the voyage to the hulks of the Thames in 1819. That was only the beginning of their journey, for all were to be transported to Australia for various sentences from seven years to the term of their natural life.

However, even when there was no enemy to worry about, voyages could have their excitements. In December 1814, the smack *Forth*, under Captain Tul-loch hit bad weather in Yarmouth Roads on her voyage back to Leith from London. She was sailing from the harbour when the wind suddenly shifted and drove her onto the bar. The crew helped the forty passengers through the wild surf to safety, and then unloaded the cargo in case *Forth* broke up. As it happened, that extra effort was not required as she was refloated and survived to complete her voyage.

Nobody remembers the smacks now, but in their day they had character and courage.

Convoys and privateers

During these wars, it was a common practice among maritime nations for private individuals to buy a licence for arming a ship to hunt down enemy vessels. These private warships were known as 'Letters of Marque' or more simply as 'privateers'. The owners hoped to recoup the cost of the license by selling the ships and cargoes they captured from the enemy.

With the Royal Navy usually dominant at sea by the early nineteenth century, Britain's enemies, whether France, Spain, the Netherlands or the United States of America, resorted to privateers and they made some devastating inroads into British shipping. Naturally, Scottish seamen retaliated, and Scotland also sent out letters-of-marque to hunt for enemy merchantmen. However, foreign privateers had a distinct advantage, as there were many more British merchant vessels than French, Spanish or American, so sometimes British ships found themselves facing heavily gunned enemy warships rather than fat and profitable traders. On other occasions, they met enemy privateers.

In the summer of 1760, the Leith privateer *Edinburgh*, commanded by Captain Thomas Murray, with eighteen four-pounder cannon and a few swivels - anti-personnel weapons for close combat - hunted for enemy vessels in the North Sea and the Atlantic. At latitude 13 degrees north, 58 west, a fourteen gun French privateer attacked her.

The Frenchman was equally eager to fight, so both vessels exchanged broadsides, manoeuvred and fired again throughout a long, hot and tiring day. The French ship was severely crippled aloft and withdrew for repairs, leaving Captain Murray to patch the damage to his vessel and continue his voyage. The Frenchman returned, and the fight continued, with the Frenchman closing to board and Captain Murray ordering the crew of *Edinburgh* to the swivels and handing out muskets to those who were not otherwise employed.

With ammunition short, Captain Murray held his fire as the Frenchman loosed broadside after broadside until their captain thought they had sufficiently softened *Edinburgh*. The Frenchman came close with her crew crowded on deck, waving boarding pikes and cutlasses, shouting their victory songs- and Murray ordered his men to fire.

The combination of anti-personal swivels and musketry was devastating, scything down the massed French crew. The Frenchman put on all her sails and fled, with *Edinburgh* in hot pursuit, but the Frenchman escaped, and *Edinburgh* put into Barbados to repair and land her wounded. There was no profit in that encounter, but seamen along the banks of the Forth spoke about it for many years.

Another Leith privateer was *Simms*, which sailed in 1806 for the Greenland Sea, hoping to catch a Dutch or French whaling ship, or perhaps an East India-man homeward bound from the Orient. The Indiamen would take the North-about route to avoid the Royal-Navy dominated English Channel. Captain Kelly took command of *Simms* as she sailed out of Leith, with eighty-five men and fourteen guns, but luck was not with her that year.

Rather than enemy whaling vessels, at 77 degrees north the French National frigate *Le Guerrier* with fifty-two cannon found *Simms*. The French ship was one of seven that had avoided the British blockade and escaped from Lorient in Brittany. Captain Huber of *Le Guerrier* had overwhelming force and subdued the resistance of the much smaller Leith vessel. After that, a horde of French-men rushed on board *Simms* and looted her from stem to stern. Kelly thought that the French crew looked very sick, possibly because the Royal Navy had blockaded them in port for so long they were not used to seafaring. Despite be-ing enemies, the French crew did not maltreat the men from *Simms*, but acted as fellow seamen throughout: the looting was expected in time of war.

As the victors herded Captain Kelly and the crew onto the French ship, *Le Guerrier* attached a tow rope and hauled *Simms* southward. *Le Guerrier* was not content with just one British prize, however, and aso captured *Boyne* of Yarmouth, Archangel bound. Huber must have been disappointed that *Boyne* was in ballast, so yielded no valuable cargo, and he ordered her burned. *La Guerrier* had already captured *Dingwall*, a London whaling ship, and *William* of Greenock, bound for Newfoundland, plus some other vessels she pounced on off the Hebrides. After looting her prizes of all that was valuable, she burned them to the water.

For ten days, the crew of *Simms* languished as prisoners, and then Captain Huber hailed a passing Danish vessel and transferred a number of them. There was another transfer when the Danish ship met an Aberdeen vessel, and the men eventually arrived in Shetland, from where they found passage to Ab-erdeen.

However, their trials were not over, for it was a long way from Aberdeen to Leith and many of the men had to beg their way south as distressed mariners. Other of the crew caught a passage with *Lord Fife*, Captain Craigie, and landed directly in Leith.

The authorities in Leith were alarmed at the thought of a French flotilla cruising in Northern waters so did something about it. Real Admiral Vashon called up the Sea Fencibles from Newhaven and Fisherrow, hardy fishermen who had volunteered as a Royal Navy reserve, manned the 64 gun guard ship *Texel*, put Captain Campbell in command and ordered her to hunt for Frenchmen. The Fife Coast Sea Fencibles and the Galloway Militia also offered their services, but *Texel* had sufficient volunteers so did not take them, much to their disappointment. Texel was not the only vessel sent north but was part of a squadron that also included three frigates and a sloop. Although *Texel* saw no action, there was no doubting the commitment of the Forth Seamen.

In September 1806 the splendidly named Association for the Defence of the Firth of Forth were so pleased with the 'alacrity and zeal' of the Sea Fencibles that they voted to pay fifty guineas to Captain David Milne, who commanded them. The zealous captain was instructed to distribute the sum among his men. It was an unusual, and no doubt welcome sign of appreciation.

Convoys

As an important naval base, Leith was a gathering place for convoys to the Baltic. For example on the 21st April 1809, two convoys left Leith Roads on the same day: one, escorted by the sloop of war *Clio* headed to Gothenburg and a second sailed to Heligoland, under the watchful eyes of *Fancy*, a gun brig. Shortly after another gun brig, *Forward* shepherded a small number of vessels to North America. Gothenburg was a common destination for convoys from the Forth, with, for example, the sloop of war *Snake* escorting thirty-four ships there in September 1812. At the mouth of the Forth, the convoy passed the gun brig, *Gallant*, returning from a cruise to locate enemy shipping. On the 26th of that same month *Peacock*, a sloop of war, sailed from Burntisland Roads with a transport and three other vessels for London, with another for Hull, homeward bound from Archangel, joining them for protection.

Leith was also the headquarters of a major smuggling operation from Britain into Bonaparte's Europe through the island of Heligoland. The vessels would sail from Leith, jink the French coast guard boats and land British goods in defiance of Bonaparte's Continental System.

The Forth was a busy place in those heady, dangerous, exciting days when Bonaparte was the enemy and convoys gathered under the watchful eyes of the navy, smacks bustled southward, wary of privateers and prizes were sold to the merchants of the Forth.

Chapter Seven

The Greenlandmen

Although people still remember the whaling industries of Dundee and Peterhead in song and story, they seldom sing about whaling from the Forth. That is a shame, for Leith was the first Scottish port to send ships to the Arctic, and there were many stories woven around the Leith ships. In 1750 Leith started a century and a half of Scottish Arctic whaling when the ship *Trial* headed north. Only the following year Captain Murray sailed to the Arctic with a secret weapon termed 'a curious machine for throwing harpoons', which a man with the evocative name of Bond invented.

The ships were known as blubber boats or Greenlandmen, with the latter title also being applied to the seamen who followed this trade. They sailed to what was then termed the Greenland Sea, around Spitsbergen, although as the years passed the focus of operations altered to the Davis Strait, between Greenland and Eastern Canada.

It is hard to put oneself in the place of these early whaling men. At that time much of the world was still a blank, Australia and New Zealand were not known, most of the African interior was unmapped, and academics disputed the existence of some mythical animals. Take the unicorn for instance. In 1754 the Leith whaling ship Prince of Wales carried home 'the entire head and horn of what was then called a 'sea unicorn', now known as a narwhal. That discovery led to a newspaper discussion that contained the following telling statement: 'in the opinion of the most knowing of the moderns, there is no such thing as a land unicorn'. Whaling men, however, were notoriously superstitious, many believing in witches, seeking fortune tellers before voyages and seeking the company of a 'lucky' captain. With this combination of superstition and sheer courage, men from Leith and other Forth ports sailed north to hunt the whales.

The industry

Some of the Leith whaling ships sported notable names: *Royal Bounty, Raith, Six Brothers, Neptune,* and *Grampus*; names that hide the hardship and horrors of the far north. The vessels would leave the Forth in late winter or early spring for the Arctic. They hunted whales and seals along the pack ice or among the bergs of the Arctic, with danger from storms or ice, fog or the vicious swing of a whale's tail.

The method was simple: the ships would carry four, five or six small boats, known as whaleboats. When they reached the whaling grounds, the area of sea and ice where the whales were most likely to be found, the whaleboats readier for immediate launching. Each boat carried six men, with three specialists, a boat steerer, line manager and harpooner, and three full-time oarsmen. The boatsteerer took them to the best position for harpooning the whale, the harpooner would throw or fire the weapon, and the line manager ensured that the line between the harpooned whale and the boat ran true without kinks.

Once harpooned, the whale would swim away, towing the boat, or perhaps a few boats until it was too tired to run. When it lay still on the surface, the whaleboats would close and finish it with lances. The dead animal was then dragged back to the ship for the harpooners to strip it of blubber, which was boiled down for oil, while the whalebone or baleen from the mouth had a hundred different uses, from brushes to corsets to netting and springs.

Whaling could be a dangerous job. As well as the traditional seaman's perils of storms and fog, there was the fear of being crushed in the ice, or being inside the whaling boat when the whale's tail lashed down upon it or being stuck in the ice for months as scurvy ravaged the crew. Augmenting these horrors was the ultimate disaster of war.

One reason for the British government's encouragement of the whaling industry was to create a body of seamen who were hardy beyond all others. The industry was so valuable that the whaling seamen – the Greenlandmen - could obtain a document known as a Protection, which meant they were immune from the Impress service, the Press Gang. That did not stop the Royal Navy from pressing the Greenlandmen as soon as they re-entered British waters on their return from the Arctic.

Whaling ships were also valuable in themselves, particularly when they were fully laden with blubber and bone, so when war came, the enemy was keen to capture them. There were many wars throughout the eighteenth century when

Britain and France were bitter rivals for trade and empire. In each one, the iced waters of the Arctic became a battleground as both sides, plus their allies, sent warships and privateers – private vessels licensed to attack the enemy's merchant vessels – to hunt the other's whaling ships.

Wives, girl friends and sisters would wait at the quayside for the ships to return from the north, but sometimes they heard bad news. In the days before telephones, radios and even the telegraph, once a ship vanished beyond the horizon it could be weeks, months or years before wives and mothers heard anything of their men. Seamen lived in a different world that could be as alien as the far side of the moon. In times of war, enemy vessels added to the natural perils but the Royal Navy also provided escorts. For example in 1782 the Dunbar whaling ships, *North Star* and *Endeavour* were escorted to the whaling grounds by *Prince of Wales*, a sixteen gun privateer that skilled hands had converted from a whaling ship. The Dunbar whaling ships seemed well adapted to taking care of themselves, with or without an escort. For example in 1797 both *Blessed Endeavour* and *East Lothian* of Dunbar were armed and sailed together for mutual support. They were off Fair Isle on their voyage north when an enemy privateer nosed closer, but the Dunbar vessels ran out their guns, and the enemy decided to hunt for easier prey. Privateers wanted easy pickings, not hard fighting.

Whaling ships from the Forth ports and Dundee gathered in sheltered bays such as Largo Bay, where a warship would shepherd them north, but sometimes press part of the crews despite protection certificates. Privateers caused concern, but the Navy was often keen to help. For example in 1762 the Leith whaling ship *Edinburgh* ran from a fourteen gun Dunkirk privateer *Duke de Broglio*, but when she entered the Forth HM sloop *Dispatch*, commanded by Captain Bertie, turned the tables, pursued the suddenly fleeing enemy for thirty hours and captured her.

In 1789 Revolution tore France apart as the lower classes rose against the aristocracy. The monarchies of Europe watched as the guillotine reduced the aristocracy one pampered head at a time, and then they declared war. In January 1793 Britain joined the anti-French alliance, and the whaling ships were once more in the front line.

The very next year the whaling ship *Raith* sailed straight into trouble. A flotilla of French vessels was hunting in the north and snapped up *Raith* and the Dundee whaling ship *Dundee*. The Royal Navy recaptured *Dundee*, and those of her crew who were still on board told a compelling tale. They said that their French prize crew had rescued a boatload of French seamen a short time before. The Frenchmen had been part of the prize crew of *Raith*. The French had put sixteen men on *Raith* and removed most of the Leith crew, but had left three Scots on board, including the mate, a Shetlander named Lyons. The French master ordered the prize crew to sail to Bergen but had not reckoned with the men he left behind. A rogue wave washed one of the surviving three of *Raith*'s crew overboard but the others waited until the French celebrated their victory by breaching *Raith*'s rum, and lay in a drunken stupor around the deck, and then they struck. Wielding the truly vicious flensing knives, Lyons drove nine French into one of the whaleboats, locked the remainder down below and took command of the ship. Lyons took her into Bergen, and eventually, *Raith* returned to Leith, and the French prisoners were marched up to the castle. Lyons gained promotion, and next voyage took command of *Raith*.

The Royal Navy also captured enemy whaling ships, as in August 1796 when HMS *Cirte* took a Dutch vessel to Leith.

But there was a downside to whaling. When the ships brought the blubber back to Leith, it had to be boiled down to create oil, which was used for domestic and industrial lighting among other things. There was one boiling house at Timber Bush, but in the 1820s and 1830s, Peter and Christopher Wood had their boiling house off Tower Street. It was a commercial necessity, but boiling whale oil produced an extremely unpleasant aroma that permeated the entire area and was known as 'Wood's Scent' or 'Wood's Scent Bottle'. Not everyone was unhappy when the Leith whaling industry collapsed in the early 1840s.

Troubles for a surgeon

The surgeon on whaling ships occupied a very privileged position. They were normally young men, either medical students or recently qualified doctors. The students had two motives for going to sea in a whaling vessel: they would gain valuable experience of actual cases, and they would earn money which would help pay for their degree. They were also reputed to have the easiest job of any onboard, dining with the master and helping with the paperwork unless there was a casualty on board. However, there were some cases where the surgeon may not have agreed.

For example in 1814 John Nicol, the surgeon of *Elbe* of Leith, did not have a happy time with Captain Young. On the homeward voyage from Davis Straits, the pair argued, and Captain Young lifted his boot and kicked the young surgeon. While such an act was quite normal between a ship's officer and a member of the crew, it was unusual to kick a surgeon, and Nicol took Young to court, where he won a few guineas in damages and much satisfaction.

Visitor from the Arctic

The whaling industry brought other excitements to the Forth as well. Only two years later, in 1816, Leith was treated to some amazing displays by a nineteen-year-old Inuit. Crowds watched as he showed his prowess with his kayak, including winning a race to Inchkeith and back against a six oared rowing boat. He was also expert in throwing his spear or 'dart', so he bisected a ship's biscuit at thirty yards range.

Some of the men watching would remember their early in whaling ships, think of the perils of the north and shudder.

Baffin Fair

Other seamen called the whaling ships 'Blubber Boat' and sneered at their ungainly appearance as they lay bluff bowed, broad- beamed and clumsy in port. Other seamen, however, did not have to endure the frightening conditions of the Arctic north where the pressure of ice would crush any ship not built with a double planked hull and triple reinforced bow. The blubber-boats were neither elegant nor fast, but perfect for a particular task. They were whale hunters, crewed by a blend of experienced Greenlandmen and the adventurous first voyagers, captained by men who understood the weft of the ice and could read the blink of skies above the freezing Polar sea.

The Greenlandmen left Leith in early spring to the scrape of the fiddle and scalp-lifting wail of the pipes. Coloured ribbons rustled in the rigging to provide slashes of gaiety for a crew half of whom were suffering from the effects of too much drink the previous night, and all of whom stared back at the packed quayside. The women, either worrying wife or broken-hearted sweetheart, waved back until the ship was only a memory.

> *'Is that father out there in the ship, mother?'*
> *'Yes my pet; he's sailing to the bitter north to find whales to hunt?'*
> *'Why is he sailing so very far away, mother?'*
> *'To earn his bread, my pet, and pay the bills for us.'*

The whaling women knew that the voyage would be arduous; they knew that there was no guarantee that they would see their men again, for it was common for Greenlandmen to die of cold, exposure or disease in the bitter chill of the north. Often an entire ship was lost, crushed in the ice or sunk beneath the Atlantic waves.

From the homeport, the ship would slog north, to Orkney or Shetland. Here they augmented the crew with some of the most skilled small boat sailors in the world while the 'southern boys' had their last shore leave for months. Many Island men signed on for extra money, but landlords denied some that right. When men from the Out Skerries joined the fleet in the 1850s, the Tacksman fined them. The sea might offer hardship, but compared to the servitude of feudalism, it was free. No wonder that men from Shetland sailed all the oceans of the world.

When the whaling ships dropped anchor off Lerwick or Stromness, the bumboats swarmed out, bartering fresh food and beautiful knitting for the manufactured goods of the cities. There was shore leave too, so the Southern Boys could carouse in the whisky bothies, or listen to the weird words of a fortune teller, or part with a silver sixpence to secure favourable winds for their ship. There was also wild dancing amidst the reek of a peat fire or a last hurried intimacy with the less moral of the island women before the long voyage where women were only a sweet memory.

'Will father think of us when he's in the cold and barren north?'

'Yes my pet,' the mother said 'he'll pray for us every day.' As I do for him, thought the wife.

The more astute of the First Voyagers would buy insurance for the grim times ahead by spending some of their month's advance wages on a bottle of whisky. To the Shetlanders this spring invasion was 'Greenland Weather'; a time of wild winds and raucous sailors, but the island men who sailed north would bring back fresh stories for the winter firesides and money for wives and children.

From the islands, it was north and west to the whaling grounds. By the second decade of the nineteenth century, most ships crossed the Atlantic to the Davis Strait, avoiding the fog and icebergs that clustered round Cape Farewell at the southern tip of Greenland. Once they reached the entrance of the Strait men were sent into the rigging or up to the Crow's Nest to look out for floating ice, particularly the 'washing pieces' that waited low under the waves, flat and wickedly dangerous in the water.

The boats were ready too; 'on bran' for the first sight of a whale, but that first sighting grew increasingly unlikely as the years passed. Ships had to sail further north, and still further, with the wind bitter-chill and the men huddled in the forecastle, waiting for the sighting of a whale, hoping for a decent catch, thinking of their wives back home.

'Mother, it's been weeks now. Will father be home soon?'

'Not yet, my little one; he still has far to go.'

And the mother comforted the child, while the wife worried about her man.

And still, they sailed north, up Greenland's west coast, the ships glistening with ice, the men cold, already tired, cramped down below or shivering up aloft.

On the 1st of May or when they breached the Arctic Circle there was a formal ceremony to the sea god Neptune. The veterans cheerfully grabbed the First Voyagers and shaved their heads before pouring tar or the contents of a piss-pot over them. Those First Voyagers who had managed to retain their Shetland whisky produced it now and escaped the worst of the abuse. The Greenlandmen poured some of the whisky into the sea as a libation to Neptune, but the whalers themselves would drink most. There was dancing to bagpipe or fiddle, and a garland of ribbons was hoisted up the main topmast, with an effigy of the ship in the centre to bring good luck for the weeks ahead.

Often this was ineffective.

For the ceremony marked only the beginning and all the work and most of the danger lay ahead.

At one time the whalers could catch their whales in the Greenland Sea or the lower reaches of the Davis Strait, but a flood of whale-hunters from Germany and France, Germany, Scotland and England had decimated the herds. By 1830 the ships had to push much further north, following the passage pioneered by Larkins of Leith in 1817. Together with a ship from Hull, Larkins had crossed Melville Bay to the 'west water' where whales were plentiful. The following year John Ross the explorer pushed even further, and oil-hungry whale ships followed in his wake.

Over a hundred miles wide, Melville Bay has an island strewn northern coast where the water is so shallow that it holds the ice well into June or even July. There is a ribbon of clear water south of this bay ice, but bergs drifting north-ward make the passage so hazardous that the whalers christened Melville Bay the 'Breaking-Up Yard.' It was not ice that broke up here, but ships; robust, ungainly, triple reinforced blubber-boats were as vulnerable as a china cup before the terrible pressure of ice.

In June of 1830, the British whaling fleet congregated at the Devil's Thumb, a prominent column of rock to which some Greenlandman had affixed a suitably sardonically name. The superstitious quietly doffed their caps to the Thumb, the God-fearing turned their head, and the ship masters took soundings and checked their equipment. The Breaking-Up Yard lay ahead; the testing time was upon them, but so was the reason for this trip, once across Melville Bay they could catch whales to make money that would pay the rent and feed the family.

That summer in 1830 there was a busy scene at the Devil's Thumb, with around one hundred ships in three distinct groups, whalers from Dundee and Leith, Montrose and Kirkcaldy, Aberdeen and Hull, tall masted swaying gently, men calling out caustic greetings to companions in rival blubber boats. So far the weather had been kind with the ice lighter than normal, so when the leading vessels saw a clear passage to the west water they did not hesitate. The veteran *Eliza Swan* of Montrose was first into the passage, followed by *St Andrew*, with a long column of ships in her wake. There were five vessels in that first group, followed by another line of four, then twenty-two including *North Pole* of Leith threading through the sliver of clear water into the heart of the Breaking-Up-Yard. The ships were so close that a biscuit dropped from the taffrail of one could land on the bowsprit of the next. There might have been a fiddler scraping a melody on one, or a piper sending a pibroch to challenge the dullness out of the Arctic air.

When they came to ice the ship's master could break through by setting his crew to rock the boat, with men running from larboard to starboard and back until the rolling motion cracked a passage clear. Or one of the stoutly built whaling boats could be positioned above the bowsprit and dropped, again to smash the ice. When the wind faded to nothing and the ship lay lifeless and becalmed, men would scramble into the ship's boats to tow her. If there was not enough clear water for even a ship's boat, they would haul her by sheer muscle-power, slithering along the ice and straining on massive lines until the sweat started from their bodies and they cursed the day they left the comfort of their homes.

With the first twenty-three ships inching along, the remainder manoeuvred to follow, but the ice closed. One minute there was open water, the next there was just ice. Divided from their comrades but united in the pursuit of the whale, the twenty-three continued onward toward the hunting grounds of the West Water. On the 19th June, the wind rose from the south west and thrust masses of floating ice toward the blubber-boats. Just south of Cape York an entire six-ship column was trapped, each ship hard against her neighbour and the wind continued to rise. Soon a full gale screamed through the rigging of the trapped ships, drove sleet hail and bitter, horizontal snow at the muffled men on deck.

> *'Mother, will father be all right? I can feel it cold tonight.'*
> *'Hush my pet; your father is a sailor. He knows what he's doing.'*
> *But the wife pulled her coat over her shoulders and shivered as she looked out to sea.*

With the sea ice pressing remorselessly and the bay ice an immovable barrier, the masters called out the hands to cut docks in the ice. Understanding the necessity, the experienced men would not complain at this muscle-numbing work as they tumbled from the shelter of the ship to stand exposed on the shifting, pressing ice. They would hack at the iron-hard surface with huge saws a length and half the length of a man, rasping and swearing, as is the way of Scottish seamen, and jesting while the sweat dripping from their faces and froze along the tangled hair of their beards. They knew what they were doing, carving out a harbour in which the ships could find shelter, and they trained the First Voyagers in the techniques of Arctic survival. Many had performed this task before, and usually, it worked, but not this time. Such was the force of the wind that the floes piled one on another, above the level of the hacked out harbours, and continued to press on the vulnerable, motionless ships. If they had not been blubber-boats, not been double hulled and reinforced with internal beams of solid oak, the ice would have crushed them long before, but even these stout vessels could not last forever.

Eliza Swan of Montrose, veteran of decades of Arctic work and the American wars, was first to be hit. A single floe scraped toward her, slammed into her hull and, raising her high, thrust the heavy whaler against the second ship in line. Ice also lifted *Saint Andrew* and crumpled a score of specially strengthened oak planks like rotten cheese, and then the floe slithered past.

Raised entirely from her harbour, *Eliza Swan* sat, damaged but repairable, on top of the ice. *Saint Andrew* was in worse condition, but still afloat. The Green-landmen set to work at once; they knew that unless they rectified the damage, much worse could occur. And as they laboured, the men knew that they had been lucky, for of the six ships in that column, only these two had escaped. The others, four stout-timbered blubber-boats, had been sunk in fifteen minutes of timber-rending horror. Despite the disaster, there were no deaths. Men spoke of the 'grinding noise of the ice tearing open their sides', of 'masts breaking off and falling in every direction' and of the 'cries of two hundred sailors leaping to the frozen surface.'

Many of the two hundred were experienced Greenlandmen; incidents such as this were an accepted part of the job, so they were prepared. Remaining at least partially dressed at all times, few men ventured far from a bundle of their clothes and possessions. Now they stood on the ice, some in their long woollen underwear, pulling on their clothes, others joking as their ship disappeared, or lighting a clay pipe with trembling fingers.

> *'And will father be home today mother?'*
> *'No my pet, nor tomorrow.'*
> *There was a chill in the air as the wife looked north. She could feel that something had happened; wives knew these things.*

Another seven ships went down before wind and ice, but for the vessels left behind at the Devil's Thumb, things were easier. They waited for another gap in the ice, a chance for them to face the Breaking-Up–Yard, but the cold retained its grip, and the ethereal beauty of the frost decorated the rigging and rimmed the garland of ribbons. Not until July did the weather change, and then it roared from the south like a hurricane. Again the ships had no sea room in which to run before the wind, little chance to face the wind head on. They had to rely on the seamanship of the Master and the will of the Lord. There was an old whaler's saying that 'No man is an atheist in a storm' and now the most foul mouthed of mariners prayed for grace as the wind drove massive chunks of ice against the wooden hulls of their ships. Without time or space to manoeuvre, the men could only watch as sabre-sharp ice sliced into their homes. Despite the double or triple strength planking, the ships had no chance. One ship's surgeon spoke of watching as a floe smashed through one bulkhead of his cabin, to meet another floe that that crashed through the opposite side. It was a nightmare come true for the owners who lost their investment and property, but a living hell for the men who tumbled into the white Hades of an Arctic storm.

Altogether nineteen ships were lost that summer season, but that was only part of the drama. There were around a thousand men stranded on the Arctic ice, shipwrecked and homeless. Added to the physical discomfort was the mental distress. They knew that without a ship they could not work, could not earn the oil money or fast-boat money that alone would ensure a comfortable winter for their family. Now they were only a useless burden.

> *'I'm hungry; is there any more to eat?'*
> *'Patience my little one. Your father will be home soon; then we can all eat our fill.'*
> *The mother prayed for a hold full of blubber even as she read the despair in the wind, so the wife pleaded only for a safe voyage home.*

If some dwelled on their bad luck, others of the Greenlandmen had other things on their mind. It was their custom to obey orders while on board but act wild as Highland heather ashore. Now they were ashore, there were piles of salvaged supplies, there were ships slowly slipping beneath the ice, and the officers no longer had authority. After all, a ship's master without a ship was only another man. The stranded Greenlandmen erected their tents, upended whaleboats to provide shelter and quickly created an encampment in this Breaking-Up-Yard. According to whaling tradition, it was acceptable, even expected to burn a wrecked ship, for that freed the stores held in the hold. As the officers watched, the Greenlandmen did just that, and the flaming wrecks burned like Viking funeral pyres, adding warmth and colour to the strange little settlement on the ice. They termed the southwest wind an 'ale wind', as their natural instinct was to rescue the liquor from the stranded ships. The Greenlandmen added ale and whisky, rum, rum and more rum to the piled up stores.

> *'Will father be all right mother, up there in the cold, cold ice?'*
> *'Your father will be working hard, my pet, working to keep us warm and fed.'*

In later years, the Greenlandmen referred to this time as the 'Baffin Fair'. For three weeks the whaling men lived on the ice, drinking and singing, drinking and dancing, drinking and playing wild melodies on the fiddle, drinking while the pipe music wailed under the bitter sky, and then drinking again. There were card games around the rum casks, and if the ship's officers gave any commands, the hands ignored them amidst the general conviviality. Although there had been no lives lost when the ships had sunk, now there were casualties. Drunken men carousing in the Arctic night fell victim to frostbite and exposure, but amazingly out of all the hundreds of possible victims, only nine or ten died.

It was the latter part of July before the wind eased, the ice relaxed its hold, and the remaining vessels of the fleet broke through. Nineteen ships had sunk, and twelve more were damaged. Twenty-one returned without catching a single whale. Leith lost two vessels, *Baffin* and *Rattler*, with another, *William and Ann* returning without catching a single whale. It had been a disastrous year for the whaling fleet, but it had been a year that entered the folklore of the Greenlandmen. At the foys of the future and around a thousand firesides from Unst to Whitby, men reminisced about the Baffin Fair of 1830.

'*And was it cold up there on the ice father?*'
'*Cold?*' *The father's eyes crinkled into a smile.* '*Cold enough to freeze your toes off, boy, but I had the memory of you and your mother to keep me warm.*'
'*And the rum,*' *his wife reminded.*
'*Aye. Maybe that too.*'

After the demise of Arctic whaling, Leith ships sailed south to the Antarctic and a host of new stories. But the Forth had plenty of its own.

Chapter Eight

The Inner Islands

Inchkeith

Inchkeith is the large island that dominates the Firth of Forth. From Edinburgh or Fife, it sits square in the centre, bulky, dark and mysterious. The name itself may come from the Gaelic *Innis Cheith*, Island of the wood, or more likely from the Keith family that once owned this strategic island. Part of the Kingdom of Fife, it boasts a lighthouse and a collection of military architecture stretching back for centuries, while people seem to have lived here, in one form or another, for around 1800 years. It is the only Forth island on which a full-scale battle took place; it has been a scene of shipwreck and quarantine, a prison and a place admired by that great Scottophobe, Dr Johnston. Yet despite its proximity to the Capital, it is unlikely that one in a thousand Edinburgh people have stepped foot on Inchkeith, and around the same number will be able to tell something of its history.

Why should that be?

Possibly because Edinburgh traditionally turns her back on the sea and faces inland, so the Forth is a backdrop, not a waterway central to the everyday lives of the citizens, and that is a real shame given the tales that attach to this island.

Geologically, Inchkeith is nothing special. It is composed of igneous rock, mingled with sandstone and outcrops of coal, shale and limestone. There are; however, fossils hiding among the shale, for those with the time and patience to search. There are also a few fresh water springs, which people probably used when they lived on this island.

Historically, Inchkeith is possibly as fascinating as Inchcolm or the Bass, which is to say that every square yard has something of interest. It is possible that Inchkeith was the base for the Otadeni township of Alauna, mentioned by Ptolemy, but if so it must have been a very small town, presumably only a fishing village or a place of refuge. It is equally possible that Bede mentioned Inchkeith when he wrote of Penda, King of Mercia forcing Osuiu of Northumberland to ransom the city of Giudi, which sat on an island. The idea of so much activity on Inchkeith is fascinating, but the stories seem more like wishful thinking than historical fact. To add to the mythological mixture is the legend that Adomnan, the Abbot of Iona met St Serf on Inchkeith and founded a religious school here.

The legends seem to merge with something more historical with the tradition that in 1010 King Malcolm II, one of mediaeval Scotland's most capable kings, handed Inchkeith to one of his followers for his exploits against the Danes at the Battle of Barry in Forfarshire. That warrior was said to be the ancestors of the Keiths. We do know that the Keiths owned the island from an early date. The legends continue with the tale that James IV, that consummate Renaissance prince, stranded a pair of infants on the island with a deaf and dumb nurse, with the intention of finding out what language they would speak. Lyndsay of Pitscottie wrote about this experiment, adding. 'Some say they spak guid Hebrew, but I know not.'

In 1497 documented history makes a welcome appearance, as Inchkeith was utilised as an isolation ward for the grandgore- syphilis - that was sweeping Edinburgh. With no known cure, and no real knowledge of the disease, quarantine on an island such as Inchkeith was the perfect answer. The fact that it was used for this purpose suggests that Inchkeith was uninhabited at this time. In 1589 the plague-carrying crew of a visiting ship was sent here, the same happened in 1609 and even as late as 1799, when a Russian vessel, Britain's ally against Revolutionary France, called at Leith. Those unfortunates who died of some now-unknown disease were buried on this island, far from their home.

An island so large, and sitting close to the main shipping channel into Leith and the busy Forth ports, was obviously a tempting target for enemy attackers. The English fleet came north in 1547, supporting their army of invasion that destroyed Edinburgh and defeated the Scots at the battle of Pinkie. The Earl of Somerset sent his men to occupy Inchkeith, where they built a square fortress complete with corner towers and garrisoned it with a company of Italian mercenaries. The nationality of the soldiers is a reminder of the international nature of much of Scotland's wars; the English used Welsh archers, Irish infantry and mercenaries from as far afield as Greece, Germany and Italy, while Germans occasionally and French frequently helped the Scots.

On the 29th June 1549, General D'Esse led a combined force of Scots and French that landed to cleanse Inchkeith, and there was bloody fighting that left some 'three or four hundred' Italian dead. Mary of Guise, the mother of Queen Mary of Scots, visited the island to gloat over the corpses of her enemies. It seems that she tried to rename the island L'isle de Dieu, but nobody took any notice and Inchkeith retained its old title.

With the Italians gone, and the English repulsed from Scotland after a war as cruel as any that the country had ever seen, the French remained, and many Scots wondered if they had harboured a Gallic cuckoo in their tartan nest. The French flag flew above Inchkeith during the period when Mary of Guise ruled the country. However the Protestants turned the tide against the Catholic French, and Scots armies joined with their Auld Enemies to expel the Auld Allies. D'Esse strengthened and enlarged the Italian's fort, but eventually, the island reverted to Scottish rule, and in 1567, with the Catholic Mary, Queen of Scots, departed to English captivity, the fort was partially destroyed. Enough remained, however, for its secondary purpose as a prison. It returned to military use, briefly, in the seventeenth century when Cromwell's army built their standard fortifications, but when they left, peace returned to Inchkeith. For the entire eighteenth-century, the island did not hear a shot fired in anger, and the noisiest visitor was the famous Dr Johnson, who claimed he would like to build a house there, for 'a rich man ... would have many visitors from Edinburgh.'

When Robert Stevenson and Robert Smith, began work on the Inchkeith Lighthouse in 1803 it seemed that Inchkeith would be a place of safety, but it remained dangerous for shipping.

Shipwreck

On 21 April 1809 the sloop *Betsey* of MacDuff, under Captain Adamson was en route from Sunderland to Peterhead but ran into a gale at Buchanness and was pushed southward, well off her route. Adamson steered her into the Forth for shelter, but things got worse when *Betsey* sprang a leak and began to settle lower into the water. The tide and wind drove her toward the guardian rocks of Inchkeith, and the panicking crew threw out the anchor and prayed to God for deliverance. However, the anchors did not hold, and the sea began to pound the ship to pieces. Adamson gave the orders for the crew to cut the anchor cables and *Betsey* ended on Inchkeith.

With the sea a maelstrom and the ship a shambles, the men shouted for help. The only people on the island were the lighthouse keeper and his wife, and of course, they left the house and ran through the driving rain to help. The struggled into the surf and managed to get the master, mate and a single seaman onto the island. Adamson had been at the helm since the storm first hit and was so exhausted that the keeper had to carry him ashore. He survived, but the ship's boy did not, while the mate lived long enough to land on Inchkeith, where he died in the arms of the lighthouse keeper's wife. In foul weather, there was always the possibility of tragedy in the Forth.

War again

With the turmoil of the French Revolution and Napoleonic Wars, the army arrived again, erecting gun batteries in case Bonaparte should sail a French fleet up the Forth. Perhaps that sounds a little far-fetched to British people brought up on the legends of Nelson's victorious Royal Navy, but in 1797 a Dutch fleet planned to land a French army near Edinburgh until Dundee-born Admiral Adam Duncan stopped them at the Battle of Camperdown. In the event, the French did not seriously threaten the Forth yet fears remained. In 1878 the Royal Engineers improved the defences with a separate gun battery on each corner of this triangular island. Once again the Forth was armed against possible invaders, and the crack of practice gunfire startled the seabirds whose natural home was this island.

The First World War saw more fortifications and strings of barbed wire stretching across the shore. Quick-fire guns and six-inch guns readied to defend Edinburgh, the Forth Bridge and the naval base of Rosyth from German invasion or attack, but save for a few scares from submarines and mines; no Germans penetrated the Forth. In 1918 an uneasy peace returned, amidst the turmoil of a disturbed Europe.

During the Second World War, Inchkeith resembled an armed camp more than an island of grass; there were Nissen huts and rocket flares in case of air raid or invasion; there was radar to watch for the Luftwaffe, Bren and Bofors guns and artillery of the Heavy Coastal Defence to fend off surface vessels. However, sometimes the guns were used for other purposes.

It was a February morning in 1940, and the gunners manning the six-inch artillery noticed an armed trawler steering directly toward a British minefield near Inchkeith. The gunners, either hoping to be helpful or so bored they grabbed any excuse to fire their guns, loosed off a warning shot, but the shell ricocheted from the sea and bounced right into Leith, smashed into a house in Salamander Street and tore a massive hole in the front of the building. If the shell had been filled with explosive there could have been a tragedy, but the gunners had used a practice shot, a wooden case filled with sand, and there were no injuries save a slight cut to the occupant, Mrs Cairns. There were no other occasions when the guns from Inchkeith bombarded Edinburgh and the hundred and sixty men from the garrison remained on the island until 1943 when the fittest were syphoned off into infantry regiments; some took part in the Normandy landings and subsequent campaign to liberate Europe.

Inchkeith played an important, and little-known, part in the success of Normandy, for the island was used during the deception plan known as Operation Bodyguard. By early 1944 the Germans were well aware that the Allies planned to invade northwest Europe. The only questions were when and exactly where. To keep them guessing, the Allies created various deceptions, including a fake plan to invade Norway with the British Fourth Army, based in Edinburgh. There was no Fourth Army, but false radio calls fooled the Germans, while double agents sent spurious messages to heighten the deception. A small unit of this non-existent Fourth Army was employed to make false attacks and landings on Inchkeith, scaling the cliffs and making as much noise as possible to hold the German's attention while the genuine preparations continued far to the south.

Overall the ruse worked, with German forces not being fully committed to the Normandy landings until the allies had fully established the beachhead. German aircraft checked out Inchkeith as late as October 1944. By the following year, it was evident that the German threat to Scotland had vanished and the army removed Inchkeith's defences. After the war, the island reverted to more peaceful use with a new foghorn installed in 1958. In the 1980s the lighthouse became fully automated, so there was again no human presence on the island.

Today the islands sits quietly in the centre of the Firth, a point of vision for residents of Edinburgh and visitors alike, but there is one curious fact about Inchkeith; at 21.75 inches (550mm) a year, the island has the lowest average rainfall in Scotland. That alone makes Inchkeith an object of envy. Perhaps Dr Johnson was correct, and it could be a fitting place for a house.

Inchcolm

If the Bass Rock is the dramatic introduction and Inchkeith is the father of the islands, then Inchcolm is the sacred isle, the Forth's answer to Iona, an island that could stand comparison to any in Scotland for history and interest. There is only one Inchcolm, and it is, arguably, the most visited of the Forth islands, and rightly so.

The name is ancient and undisputedly Gaelic. Inchcolm is derived from Innis Choluim, the island of St Columba; although there is no proof that the Celtic saint ever set foot on this rock. However, it is more than likely that other Celtic holy men were here; spreading the Gospel of Christ to what was probably an unenlightened corner of Scotland in the dark ages.

There are many legends attached to Inchcolm, and some may even be accurate; stories of Norse and English, of devious monks and mediaeval kings. It is a storied island.

Inchcolm is barely a quarter of a mile off the coast of Braefoot Bay, near Aberdour in Fife. According to legend, people once termed it Emora or Aemonia, and there are vague, unsubstantiated rumours of Roman galleys anchoring here. There is a stone roofed building that may be the cell of a Celtic hermit, although its age is uncertain, and there are surviving pieces of carved stone that certainly predate the Middle Ages. Most compelling, however, is the hog backed stone that dates from the tenth century and the legend that a defeated Norse ruler asked permission to bury his slain on this holy island. Shakespeare mentions this detail in Macbeth:

> That now Sweno, the Norwayes king
> Craves composition
> Nor would we deign him burial of his men
> Till he disbursed at Saint Colmes Ynch
> Ten thousand dollars, to our general use.

With the Norse victorious in the west and the holy island of Iona under their control, it seems that the Kings of Scots used Inchcolm as their sacred island. King Alexander I was also associated with Inchcolm, for when stormbound in the Forth in 1123, he sheltered here. A Celtic hermit gave him hospitality, and in return, the king promised God that he would establish a monastery on the island. The earliest known charter was dated 1162, and in 1235 the Augustinian monastery was elevated to an abbey. Unlike many of Scotland's mediaeval religious buildings, the ruins are reasonably well preserved, the short sea passage presumably being sufficient to prevent the vandalistic mobs of Protestant Reformers in the sixteenth century. Today Historic Scotland cares for Inchcolm Abbey, and there are day trips to view what is a remarkable institution.

The monks of Inchcolm, however, seem to have been strong characters. When one of the Mortimer family who owned nearby Aberdour Castle asked the monks to bury him on the island, the monks pondered whether they should allow such a thing, for Mortimer was a known rogue. However, they agreed and took his body on their boat from Fife until they were half way across the water, whereupon they tipped him into the sea. That is why the stretch of water between Fife and Inchcolm is still known as Mortimer's Deep. As always, there are different versions of this legend.

As an island with a peaceful population, Inchcolm was a natural target for raids by English pirates. Again there is more folklore and legend than hard fact, but the stories are vivid enough to be interesting and may even be based on truth. In 1335, when the Second War of Independence was flaring across the country, an English ship raided Inchcolm and stole the abbey's treasure, including a statue of Columba. A storm blew up and nearly wrecked the ship against the rocks of Inchkeith, but the skipper steered her to safety at Kinghorn. His men were badly frightened, believing that St Columba was angry with them, so they sailed back to Inchcolm and returned the stolen treasure. The English were back in 1384, and this time they torched the abbey, but again the weather interfered, with a sudden wind that extinguished the flames and saved the fabric of the building.

Nothing if not persistent, however, the English returned in 1547 after their victory at Pinkie. This time they were Protestant, after a desire for divorce powered Henry VIII's sudden religious transformation, so the power of a saint had little effect on them. Grabbing the island, they fortified it against any Scottish attack and from that time the island changed character. The abbey lost its prestige with the Reformation. Rather than a monastery, the authorities used the island as a prison, with Patrick Graham, Archbishop of St Andrews held here, as well as Euphemia, the mother of Alexander of the Isles.

After the storms of the sixteenth century, Inchcolm lay quiet for a while, a curiosity rather than a working island, but in the 1880s the morbid Victorians found a mystery after their own heart when they discovered a skeleton bricked into the walls of the abbey. The man was standing upright, but who he was, and why he was there, was not known. Was he an unfortunate workman, or an unfortunate prisoner, or perhaps a terrible example of the ancient Celtic practise of erecting a building on a sacrifice? It is unlikely that anybody will ever know.

The twentieth century saw renewed life on Inchcolm as warfare threatened Scotland once again. This time the menace came from Germany and Inchcolm played its part in the defence of the Forth in both the 1914-1918 and the 1939-1945 war. Strategically positioned to defend the naval base at Rosyth, Edinburgh and the Forth Bridge, Inchcolm had batteries of guns erected near the abbey, tunnels bored through the ancient hills and troops based where sandaled monks had once chanted psalms.

Now, however, the island has reverted to more peaceful times. Historic Scotland manages the abbey ruins, and seabirds and seals join the throngs of summer visitors to the Forth's unique holy island. There is a regular ferry service, *Maid of the Forth* that departs from the Hawes Pier at South Queensferry, and even in rough weather, the trip across the Firth is highly recommended, especially as there are no English pirates or Norse kings waiting on the water.

Cramond

To many denizens of north western Edinburgh, and to thousands of visitors, Cramond Island is an integral part of the Forth. It is the only Forth island that is joined to the mainland by a tidal causeway, similar to the larger and more famous Lindisfarne further south, and it is a delightful place to visit on a warm summer's day. At high tide the causeway is under several feet of water, so be careful of the tide.

Belonging to the Dalmeny Estate, Cramond Island is about one mile from the shore, just off the suburb, once an independent village, of the same name. It sits at the estuary of the Almond River, hence the name, Caer Almond, the fort of the Almond or possibly Caer Amon, the fort on the river. It is about a third of a mile long and has nineteen acres of weedy grassland. In common with most Forth islands, Cramond has a long history, with archaeological remains suggesting a Stone Age or possibly Roman occupation, which would fit in perfectly with both the name and the Roman relics that have been hauled from the River Almond.

More recently, sheep grazed on the island, and it may also have been a fishing station. There is no doubt that there were oyster beds offshore, as there were at many of the Forth islands, but overfishing and probably pollution destroyed the harvest. There are slight remains of a jetty at the North West, seaward side, which adds credence to the island's position as a place for fishermen. However, the stone farmstead in the centre of Cramond is much more recent, possibly as late as the nineteenth century although it was in use until the 1930s.

There is no record of warfare on Cramond, no tales of piracy or smuggling, and it was not until the Second World War that man brought military hardware to the island. With the threat of invasion and air raid by Hitler's armies, this essentially peaceful place became the site of gun emplacements and store houses, searchlights to seek out the Luftwaffe and barracks for the troops. These buildings left their scars in the shape of blocks of concrete, but more peaceful times have descended on Cramond. Overall, Cramond Island remains a pleasant place, whose shores brighten the seascape of this portion of the Edinburgh coastline.

Inchgarvie

One of the least regarded islands of the Forth, Inchgarvie is also one of the most observed, although few people will know its name and even fewer will realise the history over which they are passing. This small island is, literally, just under the Forth Bridge; it is the long slant of rock and stone buildings that sits beneath the cantilevers. As so often in the Forth, academics dispute the origin of the name. It may come from the Gaelic Innis Garbhach or Rough Island, but the Forth fishermen used to believe that the name came from the shoals of garvies or young herring that tended to congregate around the island. It is a nice thought, but unlikely.

It is also strange to think that people lived here until the fifteenth century. Presumably, they were either holy men, or fishermen, for the Forth was a prolific fishing ground. Either way, Inchgarvie would provide a bleak home in the winter, for until the advent of steam power it was not unknown for ships to spend many hours trying to sail or row across the Firth of Forth.

Although only a scrap of land, Inchgarvie has a fascinating history, supplemented with some folklore. For example, there is the story that shortly after

Hungus, King of the Picts, defeated the Anglian Athelstane in East Lothian; he stuck the Anglian's head on a spear head on Inchgarvie as a warning to any other invaders.

Later stories have a more secure provenance. With piracy always a possibility, in 1490 King James IV built a small fort here, around the same time as Sir Andrew Wood was repelling English raids, but it must have been an unpopular posting. However, the island had another and even more distasteful use only a few years later.

As every child learns at school, Christopher Columbus touched the Americas in 1492. However, fewer people know that one of the first imports from this brave new world was a nasty disease known then as grandgore, which seamen spread across Europe. Today it is better known as syphilis. When it first arrived in Europe, there was no known cure, so the victims were isolated, and where better to place sufferers than a small island, where their sickness could not spread to others? Accordingly, in September 1497, parliament passed an Act confining those unfortunates who caught grandgore to Inchgarvie and the larger Inchkeith. The afflicted people were escorted onto a ship at Leith and taken to the islands: 'there to remain till God provide for their health.'

With no medical treatment save isolation, most would probably slowly die, which is not always a pleasant thought for travellers on the Forth Bridge, rattling above the island. Nearly a century later, Inchgarvie was again used as an isolation island when the plague hit Edinburgh. Plagues of various types were a common occurrence until relatively modern times, bubonic plague, typhus, and typhoid were regular visitors to most towns before modern medicines, and in the nineteenth-century cholera joined the list. People regarded town life as freer than country living, but there were penalties to pay.

The small castle on Inchgarvie was also used as a prison from time to time, with the first recorded prisoner in 1519, and the last around 1671. Stuck in a cell inside a solid castle on a tiny island, the prisoners must have felt cut off from all humanity.

Although the castle was originally designed for use against pirates, it was also useful for sterner tasks. In 1547 the victorious English grabbed Inchgarvie and re-fortified it, holding the island for two years while armies rampaged around Scotland. It was defended again in the seventeenth century, with King Charles II inspecting the defences in 1651, shortly before Cromwell's army took the island, placed an English garrison here and held it against any royalist attack. One only hopes that the Forth treated the invaders as they deserved, with bitter winds and winter sleet.

After the Restoration, the defences languished for over a century, but the island was fortified again in the late 1770s when the fleet of John Paul Jones had brought panic to the Forth.

Termed 'the father of the United States Navy', Jones had been born in Scotland, worked on a slave ship but later emigrated to North America. He joined the Continental Navy soon after thirteen of the North American colonies declared their independence. In August 1779 Jones commanded three enemy vessels that sailed into the Forth. As so often, there was a threat to invade Britain, and the authorities had concentrated their defences on the south coast. The Forth was undefended by either warship or cannon.

By that time Jones already had a reputation for near-piracy, and there are tales of the upper classes of Edinburgh packing their belongings and fleeing, leaving the ordinary citizens to fend for themselves. Edinburgh Castle prepared for a siege, banks closed, and church bells clattered their warning of imminent invasion. Jones hovered off Leith and threatened Kirkcaldy until the hand of God appeared to defend the Scottish Godly. While the gentry panicked, the Reverend Shirra of Kirkcaldy took more positive action. Calling on the collective power of his congregation, he marched them to Pathhead Sands and prayed for a gale to blow this impudent invader out of the Forth.

The good Lord listened and provided a storm that blasted Jones and his three ships out to the North Sea. He still had exciting days ahead, making history in another encounter with the Royal Navy before joining the Russian Navy and fleeing that country after accusations of rape. Jones' raid was instrumental in having gun batteries erected in Leith and on Inchgarvie, but there were no more American or French attempts to force the Forth, so the cannon remained unused.

The twentieth century saw the army back, ready for the threat from Germany in both world wars. This time there were submarine defences as well, but save for an air raid on Rosyth and a German submarine at the Forth Bridge, no German is known to have penetrated so far up the Forth.

Inchgarvie, however, was not peaceful in the intervening years. In 1878 there were plans for a rail bridge across the Forth, and Thomas Bouch laid foundations on the island. The idea was good, but when Bouch's Tay Bridge collapsed in a welter of weak girders and high winds in 1879, the idea of a Forth crossing was temporarily laid aside. By 1890, however, confidence had returned, and an army of workmen was busy above the firth. Rather than being neglected, engineers extended Inchgarvie with a pier and created solid foundations for the mighty Forth Bridge. Part of the work was directed from an office on Inchgarvie, while workers crammed into the old castle. There was even official vandalism as stone from the castle was worked into the caissons of the bridge in an example of Victorian re-cycling that may raise eyebrows today.

Not many people set foot on Inchgarvie now, unless some intrepid yachtie anchors offshore, but there is a persistent rumour that a particularly nasty breed of rat populates the island. According to the legend, a party of people from Fife brought their dogs across on an anti-rat mission, but instead, the rats chased them away. Given the history of plague victims, imprisonment and Cromwell's army, perhaps the rats are a suitable inhabitant of what could be a glorious little island.

Inchmickery

Inchmickery is another of these small Forth islands that nobody thinks much about. It is just there, sitting in the centre of the Forth, unregarded and unheeded by the tens of thousands of people who pass over the bridges every week. Inchmickery is a tiny place a mile or two north of Edinburgh, measuring around 200 yards by 100 yards, yet the name, *Innis nam Bhiocaire* means Island of the Vicar, which suggests that at one time a Celtic hermit made his home there.

In common with many other islands, Inchmickery was also the site of defensive artillery during the Second World War, and a plethora of concrete buildings remain as depressing reminders of that era. In fact, they dominate the island, and until people can romanticise concrete-block hard-edged military architecture, the buildings on Inchmickery are an ugly blight on what is undoubtedly a historical island in a position of unmatched beauty.

More cheerfully, the island is now an RSPB Reserve, with Sandwich Terns, Roseate Terns and Eider Ducks to be seen, as well as the occasional seal. At one time there were oyster beds off shore too, but overfishing and pollution put paid to them.

The two rocks a short distance from the island are known as Cow, and Calf Island and passing boats are well advised to keep clear.

Chapter Nine

Fisherfolk

Throughout history, humans have caught fish. Fishermen feature strongly in the Bible, and fishing villages line both coasts of the Forth. Some are unassuming and hard working, others, such as Crail, picturesque. The coast of East Lothian has North Berwick, Cockenzie, Port Seton, Prestonpans, Fisherrow and Dunbar, from where fishermen set out and Greenlandmen joined the whaling ships. There is Newhaven, now reckoned as part of Edinburgh but once very much a fishing community. Fife has a whole string of fishing villages, such as Cellardyke, Pittenweem and St Monance, each with its traditions and expertise.

When these men were not fishing, they could turn their hands to other maritime endeavours.

Life saving

On June 15th, 1846 a sudden squall ripped up the Forth, catching many vessels unawares. One such was the small coaster *Mary* of Newburgh, heading to Inverkeithing for a cargo of coal. As the storm struck, *Mary* sprung a leak and began to heel over on her starboard side. The crew looked around and realised they were close to Inchcolm so prepared to abandon ship and swim to the island when a fishing boat hove into view and hailed them.

The master of the newcomer was Robert Linton of Newhaven, and he rescued all four men of *Mary*. The mate was unconscious, so Linton had him wrapped in the coats of his fishing crew and carried them all to Newhaven. Linton took the survivors to his house and laid the mate in bed, where he eventually made a full recovery.

Newhaven had been a fishing port since the sixteenth century, with Linton a very local name there, but it was not until 1793, according to legend that a local man named Thomas Brown found a herring shoal, and that began the Forth herring industry. Lady Nairne wrote a poem about the herring fishing that mentions not only the joys:

> *Wha'll buy my caller herrin'*
> *They're bonny fish and halesome fairin*
> *Buy my caller herrin'*
> *New drawn frae the Forth*
> *But also the cost:*
> *Wives and mothers maist despairing*
> *Ca' them lives of men*

Fishing families lived with death as a constant companion. It waited in the sudden Forth squall or the capsizing of an open boat, so the beating of the surf on the shore was the passage of death's pendulum, and each stroke of an oar could be a passage across the River Styx.

Dunbar, on the southernmost entrance to the Forth, knew all about herring as the so called silver darlings had been hunted along Scotland's south east coast since at least the thirteenth century. In 1577 there had been a thousand herring boats in and around Dunbar, local and foreign, and local folklore claimed that when the inevitable disasters occurred, the reason was that the fishermen had gone to sea on the Sabbath.

Herring fishing from Forth harbours peaked in the nineteenth and early twentieth centuries when Scotland exported much of the catch to Germany and Russia, but while not scouring the Forth for fish, the fishermen could also become pilots.

Pilots

In the early nineteenth century, many shipmasters had risen to their position by crawling 'through the hawse-hole' which meant they had started at the bottom and worked their way up through the ranks. Most were perfectly capable of commanding a ship out at sea, but when it came to the more complex business of bringing her into port, avoiding sandbanks and shoals, past islands with treacherous currents and hidden rocks, they needed the services of a local pilot. The pilots also helped guide in stricken vessels.

In the case of the Forth, many pilots were fishermen, and when times were slack, they could wait at the mouth of the Forth for incoming traffic. When vessels approached the entrance to the Forth, the fisherman would row or sail to them and offer their services to guide them past the tricky islands and into the various harbours or ports. However, sometimes the pilots were not as professional as they should have been. On the windy morning of 21st February 1801 *Mary* of Newcastle, commanded by Captain James Dickson was part of an outward bound convoy, but as they sailed out of Leith Roads, the wind increased to become a full gale. *Mary* sheltered in the lee of Inchkeith, but the Forth wind can be tricky and pushed her onto the Heriot Rock on the south side of the island. *Mary* lost her rudder and sustained other damage. As the Royal Navy ushered the other vessels of the convoy out of the Forth, Captain Dickson signalled for help.

A boat came out from Newhaven, with a pilot named James Carney in command, along with a crew of John Seaton, James Flucker, James Noble and James Logan. They looked at the damaged vessel and said that it was too dangerous to pilot her into Leith with the wind as it was, but they could take her to Burntisland. As Dickson was about to agree, Carney dropped his bombshell: the price, he said, was twenty-five guineas.

That was a colossal sum for the time, and it was no surprise that Captain Dickson turned him down flat. Rather than negotiate, Carney returned to his boat and sailed away, leaving *Mary* in a perilous position, damaged and rudderless off a rocky coast and with a fluky Forth wind threatening her.

However, there were better men than Carney in the Forth. Captain Lawrence Skene of Leith had also seen Dickson's distress signal and put out in a yawl. He agreed with Carney's conclusion that it was too dangerous to attempt to return to Leith and guided Mary safely into Burntisland. When the affair reached the ears of the magistrates of the burgh, they decided that the men in the Newhaven boat did not deserve their position and removed them from their list of pilots.

Sometimes the pilots were not lucky at all. In January 1881 the Aberdeen steamer James Hall was six miles south of West Wemyss when a look out spotted a small boat floating in the water. He shouted to the master, who halted the vessel and had somebody investigate.

When the master discovered that the boat was a Forth pilot boat with a dead body on board, he became very curious. Lifting the body on board, he sailed to Leith. Captain Grant of the Leith Police began his enquiries. Grant found out that the dead man was David Gibb, a pilot from Burntisland. He had been involved in a heated debate the previous night in Kinghorn, and his opponent had threatened to 'have it out with him', meaning to finish the argument with a fist-to-fist fight.

Gibb had walked away, and nobody saw him alive again. The boat was not his but belonged to the man with whom he had been fighting. To add to the mystery, Gibb was lying on a bed of sand and seaweed in the bows of the boat. The police discovered that somebody had knocked Gibb to the ground on Saturday night, and his attacker had threatened that 'if he didn't take care what he was about he would get far more than he had received.'

Later that night somebody in Kinghorn heard 'cries of distress' from the harbour, but when he ran there, he saw nobody. The police thought that Gibb might have fallen into the water, crawled into the boat and died there, or he had been attacked again and had fallen, or somebody had pushed him into the boat. There was a large cut on his forehead that argued he had fallen at one time. Despite Captain Grant's efforts, he did not solve the mystery.

While the fishermen worked at sea, their wives and daughters laboured on land, collecting mussels in the bitter dark of the pre-dawn, baiting the long lines where sometimes as many as fifteen hundred hooks waited, hawking the catch around the villages and towns of the Forth, or following the herring. The women worked every bit as hard on land as their men folk did at sea, gutting, salting and packing the catch at incredible speed as they followed the herring fleets from Lerwick to Yarmouth, Ireland to the Western Isles. Working long hours in often inclement weather, young women enjoyed the company of other women, escaped the often close confinement of their village and a taste of freedom and travel. It was also a chance to meet a handsome young fisherman outside their usual social orbit.

At home, the women controlled the purse strings and managed the family finances. Local lore stated that no fisherman was complete until he had a wife; the family was an economic unit, as well as a social necessity. Fishing families from both sides of the Forth, were close-knit and hardy. They had to be, given the nature of the work and death constantly waiting in the wind.

'Nane o' us stands by himsel!'

If the Bayeux Tapestry depicts the essence of warlike Normans set on their conquest of England, the tapestry of Eyemouth is a lasting tribute to the pride and sorrow of the East Coast fishing communities. Sewn by twenty-four expert needlewomen, the fifteen-foot tapestry depicts scenes and symbols of the danger that haunted every moment of the fisherman's life: an unexpected storm. That of the 14th October 1881 was so shocking in its suddenness and so awesome in its force that people still remember it today. It took the lives of one hundred and eighty-nine fishermen, all but sixty from Eyemouth, a small town that sits just a few miles south of the Firth of Forth.

In the happy days of plentiful fish, Scottish coastal villages thrived; fleets of brown sailed boats left their havens to haul in the nets. The fishing industry, both deep sea and inshore, created jobs for tens of thousands of people, from the fishermen who sailed the boats to the hard grafting women who gutted and salted the catch. But the seas around Scotland are notoriously dangerous, liable to sudden storms, backed by coasts of uncaring cliffs, sharp fanged rocks, unforeseen currents and rugged islands. Hardly a year passed without the sea claiming boats and men.

The fishing community accepted the sorrow and loss and pain, as they accepted the terrible toil of the women. On shore, the women were in charge, but the men belonged to the sea. And all were in thrall to the weather.

In 1881 a miserably wet summer was followed by a stormy autumn, but the men were desperate to get to sea. After seven days of continuous gales, on the evening of Thursday 13th October, the weather moderated. Fishing was a livelihood, not a job; seven days stormbound in harbour meant a week without money, hunger and hardship for the family, but even so, many of the wives were doubtful as they baited the lines.

A calm morning helped clear their fears, and on Friday the fishing fleet slipped out from the harbours and coves of the East Coast from the Border to the Neuk. They sailed from Burnmouth and Eyemouth, from Coldingham Shore and Dunbar, from Newhaven and Cellardyke, Crail and all the rest. Many of the Forth boats had left at midnight, gliding out of the harbour wall in the chill autumn dark as they sailed for either the inshore grounds near the Isle of May or the deep waters further out.

'It's a grand day' the fishwives of Eyemouth said, cheerfully relieved.

As was their habit, the fishermen checked the public weatherglass on the pier. 'Aye' they replied ', but the glass is low.' The needle pointed down; the barometer had fallen an inch overnight, but the sea was quiet in the bay.

At eight in the morning, the fleets sailed, with the Newhaven boats easing into the Forth and the Eyemouth vessels sailing around the Huscar Rocks at the harbour mouth and out of the cliff enclosed bay to the fishing grounds eight miles offshore. Although the weather remained calm, sea nearly flat, and air breathing lightly, all along the coast, many fishwives, their perceptions tuned by years of experience, felt that something was wrong. Living close to nature seems to heighten senses dulled by city life and by ten in the morning the Newhaven women were in their doorways, watching. There was nothing unusual to see; the Forth was quiet, ruffled only by a whisper of wind and the brown sailed boats were clearly visible.

Then even the slight breeze lapsed. Quiet fell on the sea. The colours altered, darkened as silence pressed ominously down. Knowing that something was wrong, many wives left their doorways and ran to the harbour, worried, hoping that their men were safe, that they would return.

'Look!' A woman pointed out to the Forth; a sudden gust of wind had erupted from the north and west to rip into the quiet, tearing chunks from the sea and raising massive waves. The women on shore could see that what approached was not an ordinary squall, but something like a hurricane. Thick clouds overhead killed the light, and at the first onslaught, many of the fishing boats were shattered – literally. Boats were thrown up out of the water or were tossed upside down or had their masts wrenched from them, or had sails shredded, so they hung in rags. A shockingly powerful wind drove the boats onto the skerries and rocks of the coast. If they survived the first shock, the fishermen either made for the harbour or struggled to keep afloat, fighting through a sea that had erupted into a screaming maelstrom.

The first picture in the Eyemouth Tapestry is a fishing boat in distress. Two men in oilskins struggle with the sea as a wave surges into the helpless boat. It is an evocative portrait, frightening, distressing.

At that time the fishing fleet was in the process of change. For decades, perhaps centuries the fishermen claimed that it was cheaper to build open boats than decked. They also thought that open boats carried more fish than those with a deck. Naturally conservative, fishermen did not like change. Yet there was a change coming to the fisheries. After a terrible storm in 1848, larger, safer boats were creeping in. From the Moray Firth came the sea-kindly Scaffies, fully decked, able to sail further out to sea and hold a larger catch. From 1879 came the lug-rigged Zulu, the brainchild of William Campbell of Lossiemouth. There was also the Fifie, a typically unassuming Forth boat with a near vertical stem, no boom and low bulwark. The fishermen of 1881, bearded, oil skinned hardy men were mainly in smaller, open yawls that were common from Berwick-upon-Tweed to Lerwick in Shetland, but even the larger boats died that day.

Watching the tortured seas, the women realised that they were witnessing something phenomenal. They cried openly with fear for their men while the children at school in Eyemouth screamed with panic until they were released to run home.

'We'll never see our faither again!' One boy sobbed. Probably his mother tried to comfort him, but she would know the truth; fishing families lived with the constant presence of death. In Eyemouth, women and children rushed to the cliffs and the seawall, flinching as powerful waves cast spray and spindrift like a high dark curtain that cut visibility to five hundred feet.

It is hard to decide who was in the worst position, the wives who stared at the terrifying seas, knowing that her husband, brother, or son was out there beyond sight; or to be aboard the boat, struggling to reach shore. The folk of Fisherrow could do nothing but watch as *Alice* headed for the harbour, sails tattered beyond belief, crew struggling with the huge waves. There would be some hope, cries of encouragement then screams of despair as the sea took *Alice.* There can be little worse in the world than to watch a husband or son die almost within reach of safety.

The storm hit the Berwickshire boats hardest. Three Coldingham yawls arrived home first, their safe arrival raising hopes. The scenes of relief when wives greeted husbands can be imagined. And when the larger *Onward* struggled into harbour things might not have seemed so bad; wives and mothers would allow themselves some optimism. At least until Andrew Dougal, skipper of *Onward* came ashore to tell his wife the news. Their son, Alexander, had been swept away only two miles offshore. *Britannia* came into Eyemouth next, and *Alabama* weathered the rocks, followed by two of the Coldingham Shore vessels. The fleet seemed to be returning, albeit in dribs and drabs, but out at sea conditions were terrible.

In the Firth of Forth around twenty boats made Burntisland in safety but others were less fortunate. Those with time to react had put out sea anchors, or lowered their sails and turned to ride before the wind. 'I never saw my boat go so fast with sail' one fisherman said 'as she did that afternoon without any.'

The hideous gale shredded dozens of sails, and the boats were driven forward with bare masts bending and the crew holding on for their lives.

A Newhaven skipper saw a boat capsized by a huge sea, with no survivors. He 'put the head of the boat to the sea and bore away, having tied the laddie to the mast.' That way the boy could not join the unfortunate Alexander Dougal. The same skipper came across another boat, keel up and no crew:

'We lifted the ballast to the front of the boat so as to put her in a better position to meet the waves and decided to have the sea for our friend for the night rather than venture near the land.'

At that period many fishermen were deeply religious, and this Newhaven skipper sang hymns as he remained at the helm for twenty-four hours. Like many other boats that rode out the storm at sea, he got her home safely.

'Waves were mountain high' another man said, adding 'three times were thrown on our beam ends.' Again this boat kept out to sea and survived.

Harmony came into Eyemouth Bay, and the anxious crowd watched, helpless, as she tried to get to windward, failed, and the storm drove her onto the rocks at the harbour mouth. *Radiant* was next, sunk just off the pier head with the onlookers frantic to help, unable to do anything. Daughters and mothers and wives heard their men shouting through the shielding spray, heard them drown amongst the rocks. The horror of that is hard to comprehend.

There were attempts to help, desperate attempts. But a storm so powerful it uprooted trees and blasted them along, upright, so powerful that it tore boats to pieces, flicked slates from roofs and created devastation from Orkney to Kent, was too much for the technology of the time. Rockets proved useless, while the wind contemptuously tossed back ropes weighed with stones. A picket boat from Newhaven bravely thrashed out in a lull, hoping to find survivors; the sea took her and her four-man crew. Unknown and unidentified ships were cast ashore all along the coast. By the evening of Friday, some battered boats had returned. Others remained out, fighting the storm, and hundreds of women hoped that their men were still afloat, still alive. Next day the search began, for survivors at sea, for bodies along the shore.

Searchers found boats, empty, upturned or with the drowned crew caught in the rigging. The sea cast bodies of men onto the shore, some stripped naked by the sea. Fishermen used boathooks to haul the bodies of friends, relatives, and colleagues to land. An occasional boat managed to limp home, sea battered but safe, with exhausted crews to give some solace in a dreadful time. Eager hands carried one skipper ashore, for, after more than forty hours at the helm with his legs twisted beneath him, gripping him in place, he was unable to stand.

The East Coast villages were used to tragedy but not on such a scale, and there was terrible grief as the women laid out the recovered bodies in their coffins for a last farewell. Not surprisingly, some women never recovered from the mental anguish of that terrible October storm. With fishing being a family business, the entire manhood of a house could have died, and with much intermarriage in the villages, everybody knew everybody else. Nobody was untouched by tragedy.

But even that had a positive side: there was a saying in the fishing communities:

'Nane o' us stands by himsel!'

And perhaps that community spirit helped those beleaguered fishing villages to survive.

Eyemouth lost 129 men out of a total East Coast loss of 189 and the Tapestry records each name reverently. But perhaps as moving is the stone monument that stands in the graveyard overlooking the sea, with the anchor proud and Eyemouth Bay a few yards distant. These men died over a century ago but the fishing communities still remember them, and still the proud boats leave Eyemouth for the fishing, passing the Huscar rocks as they go.

That storm did not end the industry. The men were too resilient to leave the sea. There is still fishing from the Forth, from Dunbar and Eyemouth. Until there are no fish left and humanity's pollution has finally destroyed the sea, there will always be fishing.

Chapter Ten

The Great Tea Race

In this twenty-first century, people purchase tea in little pouches, perfectly perforated and packaged in colourful boxes to help smooth the pain of parting with money. Supermarkets stock their shelves with dozens of different varieties of tea, and it is the work of a minute to lift the packet from a shelf and drop it into a trolley. So it is hard to believe that in the mid nineteenth-century men were prepared to race half way across the globe to arrive in port with the earliest cargo.

The latter decades of that century were the age of the clipper, arguably the best loved of all sailing ships. The very name clipper summons a vision of a sleek vessel slicing through the sea, spray slashing from narrow bows while up aloft each sail draws to the full. For once folk memory has achieved something close to accuracy, for this was very much the reality of sailing in a clipper. They were the ultimate sailing ships of their period, the maritime world spoke of their feats of speed and endurance and recalled their names with reverence. Men mentioned their captains with awe, envy, or a hint of fear. There was no greater accolade than to be the captain of a crack clipper, no greater skill than to master one of the world's fastest ships. Clippers were hard to sail, awkward to control in following seas, extraordinarily sensitive and so fine-lined that some could sail backwards up the Shanghai River. Only experts could handle them, and their masters hand-picked the best men for his crew.

It was the Opium War of 1839 – 42 that opened a handful of Chinese 'Treaty Ports' to the tea traders, and competition was fierce to be first home with the tea. In the early years the American ships were supreme, but although these ships were fast, their softwood was porous, so after a few voyages, they were strained. Scottish hardwood ships were stronger, drier and lasted longer. When the Scottish shipbuilding yards adopted iron and composite vessels, the Americans, stuck in the bloody mire of their civil war, trailed behind and lost. With the clippers of the 1860s and 1870s, the beauty of sail reached its peak. Given favourable conditions of open sea and a following wind, the best clippers were faster than any contemporary steamship, although the steamers scored with their ability to maintain a constant speed whatever the wind direction. However, the clippers redressed the balance by using God's free gift of the wind; they did not need to carry coal and did not have to call at coaling stations to refuel.

The tea trade demanded more from its captains than just superlative seamanship; it demanded daring combined with the nerve to steer a delicate ship at high speeds through both semi-charted seas and tropical storms. When put to the test, most sea masters proved too rash, too tentative or too fond of the bottle. Of the men who were staunch enough to race their crack ships thousands of miles and back, year after year, three of the best were Captain John Melville Keay of *Ariel*, Captain McKinnon of *Taeping* and Captain Innes of *Serica*. While all three were Scots, Keay was from Anstruther and the ship McKinnon commanded had been built for Captain Alexander Rodger of neighbouring Cellardyke. Both Keay and Rodger are still remembered in their home ports. All three skippers commanded Scottish built ships, and in 1866 all three competed in what was arguably the greatest of all the tea races. These races were one of the major sporting events of the era, and anyone with any interest in the sea watched for the arrival of the first China tea clipper. Owners, captains and crew wagered large sums on the result.

As April drew to a close, the ships gathered at Pagoda Anchorage, downstream from Foochow –now Fuzhou – on the Min Jiang River. There were sixteen ships and with freight for the best as high as £7 a ton; tension was high. By the time the first sampans of tea crept down the Min River the clippers had been painted and varnished, black hulls contrasting with gleaming white woodwork, brass glistening in the sun and the teak decks highly varnished. These were quality ships, and their masters ensured they looked the part, however, weather-battered they would be in the weeks of stern sailing that lay ahead.

There was *Ariel*, so sleek that even a breath of wind could slide her through the seas and a stiff breeze could spank her along at sixteen knots. *Ariel* was 195 feet long and William Steele of Greenock designed her. Steele was one of the greatest clipper designers of his time. Launched only the previous year, *Ariel* was a splendid ship and the favourite to win the race.

But she had strong competition. *Fiery Cross*, Liverpool built and commanded by Captain Robinson; *Serica* and *Taeping,* two more of Steele's ships with teak woodwork, shining stanchions, capstans and binnacles. Then there was *Taitsing* from the Glasgow yard of Charles Connell, *Falcon*, once captained by Keay, *Flying Cloud*, Aberdeen built and owned by the Hong Kong Scots Jardine Matheson, respectable business men who were alleged to have started their careers as opium smugglers. Other excellent ships included *Black Prince* and *Coulnakyle.* The competitors were ready, the crews eager for the race, all they required was the cargo and the open sea; they had the tall ships, and the skies were full of stars with which to steer them.

The first sampan carried green or old tea that was packed directly onto the ballast, as a shield for the better quality leaf that came next. Loading continued by night and day with the Chinese coolies packing expertly, efficiently, patiently. They were the unknown and unsung heroes of the tea trade but harassed mates supervised the work, ensuring the maximum volume of tea in the minimum space while always aware of the passage of time, and the demands of the ship.

On the 28th May 1866 the leading ships had been packed and on the 30th *Ariel*, with over a million pounds weight of tea on board, dropped her pilot at the mouth of the Min and headed for London. Although she had been first to cast anchor, she was not in the lead, for *Fiery Cross*, twelve hours behind at Pagoda, had whistled for a steam tug and was already ahead, with her beautiful bows dipping to the song of the sea. *Taeping* and *Serica* were almost level with *Ariel*, with *Taitsing* thirty-six hours astern. It was evident that the race would be between these five vessels. Captain Keay of Anstruther and *Ariel* was strict with his men, but harder on himself as his description of the run down the China coast illustrates:

My habit during these three weeks was never to undress except for my morning bath, and that often took the place of sleep. The naps I had were on the briefest and were mostly on deck.

Keay was not alone in this; the best captains barely left the deck during the entire three-month race, sleeping in a deck chair under the night sky and eating only when forced by hunger. The China Seas provided an opportunity to display ingenuity. Each captain had a favoured route that he was confident suited the character of his ship. But 1866 proved a wild year, and once the leading ships surged through the Formosa Channel, they were met by the Southwest monsoon. As frantic winds scudded them down the coast of Indochina, *Fiery Cross* lost some of her lead and at times the four leaders were in visual contact.

In the nineteenth century, competition was almost a religion; ferries raced to be first to port, rival railway companies forced their workers to batter across hideous terrain, business men competed for profit, and the devil take the hindmost. Success was everything, losing was frowned on and suicides commonplace among those who failed to make the grade. The tea races were only part of a period obsession.

The clippers raced on; across to Borneo and tacking along the mysterious jungle coast, catching the night land breeze and making the Api Passage to head for the Gaspar Strait and Anjer Point in Java. These were pirate waters, and the captains would be alert, thankful for the wind that blasted them along, for only in calm weather could the local vessels catch a clipper. Everybody was aware that it was only twenty years since Raja Brooke tamed the Sea-Dyaks, but wild men remained in Mindanao and predators haunted the Straits of Malacca. Other clippers had vanished in these seas so the crew would be wary as the tropic sun raised bubbles in the pitch between the deck planks and only the sea breezes cooled the sweat that burst from toiling bodies.

Twenty-one days out from Pagoda, *Fiery Cross*, *Taeping* and *Ariel* were within twenty-four hours of each other, with *Serica* only two days behind. Weather conditions were complicated, driving winds and sudden squalls breaking the tropical heat as the lovely vessels slid through the seas, always watchful for rocks and hidden reefs. If the captains had to be skilled and daring, ensuring their ships sat evenly in the water, ensuring that every sail possible was drawing, the crew was hard working. These ships were not over-manned; Ariel carried only twenty-four men, and each knew his way from the rudder to the skysail, from jib boom to taffrail, in pitch dark and howling gale. Hard men indeed, living a hard life. After Anjer Point, the Southeast Trades took the clippers clean to Mauritius, away from the Spice Islands with their enchantments and perils. On this stretch, the ships could show their best speeds with daily journeys of three hundred miles. But they were too evenly matched for any change in their relative positions.

From Mauritius, the clippers crossed to the Cape of Good Hope where the Agulhas current kicked up some heavy seas, and *Ariel* gained a little on Fiery

Cross while *Taeping* and *Serica* lagged behind. By now *Taitsing* was five days slower than *Serica* and seemingly out of the race. *Fiery Cross* and *Ariel* were in sight of each other and nearly equal, but the position changed as they rounded the Cape and surged northward through the Atlantic. The crews must have been relieved to be on the homeward leg, but there were still thousands of sea miles before they reached home waters.

As they approached lonely St Helena, *Taeping* had taken the lead from *Fiery Cross* with *Ariel* third and *Taitsing*, surprisingly, closing. The tension would be tangible, as the crew of both *Fiery Cross* and *Serica* had wagered a full month's pay on victory for their ship. Past the island with its memories of Bonaparte, and northward still with *Taeping* maintaining her narrow lead from *Fiery Cross* and *Ariel* close behind; all three crossed the Equator on the same day and when they reached the Cape Verde Islands *Ariel* was in front. *Taeping* and *Fiery Cross* sailed nearly side by side until the 17th August when the light wind changed; by a fluke, *Taeping* gained a breeze while *Fiery Cross* did not. *Taeping* pulled ahead, to the delight of her captain and crew.

In a race of this length, luck tends to even out and as they approached the Azores the leading four ships were close enough to exchange signals. It was 25th August, and *Taitsing* was now only four days behind.

At half past one in the morning of the 5th September, *Ariel* sighted the Bishops Rock Lighthouse on the Scilly Islands. Perhaps they were nearly home, but these islands had seen the death of many beautiful ships so there would be no relaxing. Captain Keay set all sail he could and sped for the Channel. With spindrift and spray forming an opaque curtain around her, Ariel tore eastward, but dawn revealed another ship on the starboard bow: the indefatigable Taeping.

The wind gusted west-southwest as *Ariel* and *Taeping* thrashed along at thirteen, sometimes fourteen knots. At eight in the morning, the Lizard light was abeam, at noon both ships were off Start Point. Six hours later they shortened sail, nearly simultaneously, off Portland Bill and by midnight they were abeam of Beachy Head.

Ariel reached Dungeness slightly ahead, and as she signalled for a pilot, Captain Keay crossed the bows of *Taeping* to ensure McKinnon could not draw level.

Just before six in the morning the pilot boarded *Ariel* and took her to the East India Dock, but *Taeping* had found a faster tug and docked twenty minutes ahead – and *Serica* was only thirty minutes astern. After a voyage of some sixteen thousand miles, all three ships arrived home the same morning. There was a bonus for the first tea, but after all the tension, all the storms and labour and drama, Captain McKinnon sportingly split this with Captain Keay and both watched Innes dock *Serica*.

Perhaps the excitement was the real reward or the knowledge of doing a good job or just displaying superb seamanship. Either way, the clipper captains were men apart. *Fiery Cross* came fourth, *Taetsing* fifth and when the race finished the captains could recover from the nervous tension. Captain Innes chose to drink a cup of tea, but he could not lift the cup without his hands shaking and spilling the contents.

Although the Clyde built *Cutty Sark* gained more fame than either of the winners of the tea race, these two leading clippers were possibly her equal, and both had a strong Forth connection. As usual, there is no song-and-dance in the Forth over the men they produced. Life goes on and the time of the clipper was short. Other methods of transport were creeping in.

Chapter Eleven

Magnificent Men in their Flying Machines

By June 1824, hot-air balloons were not new. The Montgolfiers had flown one as far back as 1783, but in the 1820s they were still enough of a novelty to attract crowds. On 28 June 1824, an English pastry chef, chemist and ardent balloonist named James Sadler prepared his balloon in the green outside George Heriot's Hospital in Edinburgh.

Sadler was a minor celebrity, but he was not the first to try flying above the Forth. In October 1785 an Italian named Vincenzo Lunardi flew from Edinburgh to Fife. Lunardi tried a second time and ended up ditching in the Forth. A local fishing boat picked him up off the Isle of May. The flying Italian was not Scotland's first aviator: that honour went to the Scotsman James Tyler, who flew Britain's first balloon over Edinburgh in 1784, only a year after Montgolfier. But James Sadler aimed to go one better: he planned to cross the Forth without ending in the water.

That June day in 1824 began grey and blustery but cleared up to be bright as Sadler piped coal gas from the Grassmarket to fill the thirty-four-foot high cone of red and white silk. Coal gas was preferred as it was cheaper than helium. After hours of preparation, it was about three in the afternoon when the hospital hoisted a Union flag to signify that the balloon was about to take off. The car was made of basketwork and shaped like a boat, lined with crimson silk inside and coloured blue outside, with a seat at either end. Sadler was not alone for Campbell of Saddell joined him. Rumours said that Campbell had paid a large sum of money for the privilege of dangling beneath a thin bag of gas, hundreds of feet above Edinburgh and the Forth. Campbell was not a slim man, and the balloon wobbled as he took his place, but still, the band played 'God save the King' as Sadler worked his magic and the balloon rose.

The crowd on Heriot's Green watched in awe, as did the scores, if not hundreds of people who stood at the upper windows of the Edinburgh tenements and even on the steep, blue slated roofs, high above the cobbled streets but soon far below the soaring balloon. There was a charge of sixpence to stand on Castle Hill and watch the ascent, but the Edinburgh crowd kicked down the makeshift fence and rushed up to watch the balloon for free. There were so many people crowded there that one man fainted in the crush. Others packed the summit of Arthur's Seat, Calton Hill and Salisbury Crags, while hundreds of excited people thronged the battlements of Edinburgh Castle.

Sadler's black cap and jacket were prominent as he looked over the edge of his basket and waved a small flag to the spectators. Opposite him in the basket - known as the cab - Campbell stood and alternatively bowed and lifted a hand; taking off his tall hat in an expansive salute to what may have been his admirers.

The wind played its own game, driving the balloon westward and then over the Old Town, where it began to descend dangerously close to the spires and chimney pots until Sadler threw down some of the bags of sand that acted as ballast. Hopefully, he checked the ground beneath to ensure nobody was standing there, but there are no reports of casualties, so presumably the bags landed in a safe spot. With the excess weight removed the balloon rose again, and the wind took it south, toward the Pentland Ridge, before deciding that north was best.

Sadler and his corpulent guest hung above the ships and docks of Leith. The watchers wondered what was happening as the balloon descended to hover above the green hump of Inchkeith before drifting over the Forth. It was so close to the water that Sadler could talk to the crew of a boat, which followed their progress in case they came down in the water.

Sadler threw a bag or two of ballast over the side, and the balloon rose again, with the wind pushing it eastward toward the entrance of the firth. People on Inchkeith cheered them and fired a gun in salute, while in return Sadler and Campbell toasted them with fine French wine. They headed northward to Elie Bay, passing some boats, each of which fired a salute, and then they dropped their anchor, an iron grappling hook, in a field at Bankhead Farm. The hook dangled above the ground until a young lady, Ann Balfour, daughter of the farmer, dashed forward, grabbed the hook despite the warnings of Sadler and Campbell that it was dangerous, and eased them to land. The balloon was back on the ground after two hours airborne.

The balloonists deflated their balloon and loaded it in a chaise-and-four for the drive to Pettycur and then the ferry to Newhaven and back into Edinburgh. It should be expected that the Forth featured in early balloon flights, but there was also a connection with later and much longer adventures in the air.

Crossing the Pond

R 34 was an airship. The concept never quite took on in the fashion of heavier than air craft, but in their time dirigibles, as they were also known, rose to brief prominence before the destruction of Hindenburg and other disasters led to a loss of public confidence. Naturally, Scotland had her share in the airship period, and again the Forth was involved.

The history of R 34's design was firmly rooted in the First World War. The Germans used their airships, Zeppelins, to bomb various towns in Britain, including Edinburgh. In 1916 a Zeppelin was shot down over Britain and captured intact. Obviously pleased at this opportunity to examine enemy technology, British scientists and engineers clustered to inspect the construction. The result of their work was a new design of British airships, named R33 and R34. The first of these was built in Yorkshire in England and the second in Renfrewshire, Scotland.

It was March 1919 before R34 flew, and after test flights, she was based at East Fortune in East Lothian, just a few miles south of the Firth of Forth. She made a flight over the Baltic in June that year, after which it was decided to go for the big one and fly her across the Atlantic. At that period long distance flights were in fashion, with nations trying to outdo each other by the first or the fastest flights over various landmasses or bodies of water. Crossing the Atlantic was a natural target.

However, R34 was not the most comfortable of craft to travel in, with passenger accommodation consisting of hammocks swaying along the keel walkway while cooking was at best rudimentary. Comfort and food were deemed irrelevant; success was the thing that mattered.

On 2 July 1919 Major George Scott gave the order, and R34 took off from East Fortune, carrying her crew, two United States airmen as observers, one stowaway and a kitten called Woopsie: the first feline to fly over the Atlantic

Ocean. There was no excitement and no adventures on the 108-hour long flight, but when they touched down at Mineola, Long Island, they were scraping the barrel for fuel. One of the travellers, Major John Pritchard, jumped out of R34 and landed by parachute to help the Americans land the airship. Pritchard was the first man to land in the United States after an air crossing from Europe. He was an interesting man, with Welsh ancestry and a father who had fought in the American Civil War but Pritchard himself had been born in England. He was unfortunately killed during test flights of R38.

R34 returned to Britain with a seventy-five-hour flight, but she did not remain in service long due to technical difficulties. All the same her flight put the Forth on the world map of aeronautical firsts, although very few people remember her today.

Chapter Twelve

Crossing the Pond

The Forth does not trumpet the ships she built, and that is a shame, for although they may not rival the Clyde in quantity and size, they certainly do in quality and significance. One of the most neglected is the Leith built *Sirius*, a 703-ton side- wheeler that was created to operate in the choppy Irish Sea but which was the first vessel to cross the Atlantic entirely under steam.

Ever since the introduction of steam power to the sea, rival shipping companies tried to outdo each other with a succession of firsts. The first ship to sail on open water, first to cross the Channel, first to cross the Irish Sea and then the North Sea and finally, the big prize, the first ship to cross the Atlantic, that maritime highway between the Old World and the New, entirely under steam power.

The Americans pushed to attempt this feat. In 1819 the New York built *Savannah* sailed into the long swells of the Atlantic with her ninety horse power engine and collapsible paddle wheels. She left from Georgia and plodded across to Liverpool in twenty-nine gruelling days. That was a historic feat, but although she certainly was the first steam ship to cross the Pond, people pointed accusing fingers and said that she had used her sails most of the voyage and had her engine running for a mere eighty hours.

The proponents of steam power were not impressed; the opponents were jubilant. Sails were still the dominant means of propulsion at sea. Steam had not yet proved itself on a long voyage. That may have been the case for *Savannah*, for when she returned to the States, her engines were removed, and she reverted entirely to sail. Despite that minor detail, the United States celebrates the days she left Georgia, 22nd May, as National Maritime Day.

The Royal Navy was next to step into the Atlantic limelight when HMS *Rising Star* made history as the first steam warship to cross the Atlantic, but again she was mainly under sail. *Royal William* was another contender; she sailed from Quebec to the Isle of Wight, but although she used her engine more often than *Savannah*, she also used her sails.

The Pretenders had thrown their dice and lost. The crown lay unclaimed, but rivals were reaching out their hands for the historic honour of being the first to cross the Atlantic under steam. The British and American Steam Navigation Company were hopeful of snatching the crown, but the Great Western Steamship Company bulldozed their contender into prominence.

The Great Western Company ordered a purpose built steamship, especially for the Atlantic crossing. In accordance with the Great Western's policy in doing everything on a grand scale, their ship, SS *Great Western* was then the largest passenger ship in the world and built to impress. Oak-hulled and strengthened with iron, she was constructed at Bristol by a man known as Bristol's greatest shipwright. Patterson was an Arbroath man who had already built Velox, a clipper type ship that turned heads in the nautical world. When the Great Western Company decided to build a ship to 'extend their railway' across the Atlantic, they asked Isambard Kingdom Brunel to design the ship but Patterson to do the actual building.

Patterson built *Great Western* in Bristol, and at 236 feet long and 1320 tons she was huge for her time. Brunel specifically went for size as he believed that larger vessels were more fuel economical. Built of oak strengthened with iron, she had four masts and first kissed the sea on the 19th July 1837. When her 750 horse power engines were added in London, she was ready for the fray. On the 31st March 1838, she sailed to Avonmouth to cross to New York, but there was a major hitch when there was a fire on board, and Brunel fell and injured himself. Many passengers seemed to think this a bad omen and refused to sail, so for all her great size, *Great Western* had only a handful of paying passengers when she finally left Britain.

In the meantime, the British and American company was experiencing problems. The company had ordered a ship built for the Atlantic crossing, but the company could not meet the deadline, so British and American had to charter *Sirius* as a makeshift.

Sirius was far smaller at 178 feet and had only a 500 horse power engine that managed a maximum speed of twelve knots on a calm day. Her designers had intended her for the confines of the Irish Sea and not the expanse of the Atlantic. Even more important, she had nothing like the fuel capacity required for the crossing. To compensate, much of the passenger accommodation was ripped out and replaced with coal bunkers. *Sirius* had one more disadvantage; her engine was of modern design with a fuel condenser that used more fuel, which was unfortunate given the length of the voyage ahead.

Make-shirt, small, hacked about and neglected by the public that lauded the size and majesty of *Great Western, Sirius* left London four days earlier, topped up her bunkers at Cork and on the 4th April 1838 she chugged into the Atlantic with forty passengers on board. The vessel built in Leith for the Irish Sea was challenging the purpose built giant on the broad Atlantic.

The race was on, with *Sirius* having the early lead but all the advantages with *Great Western*; she was larger, with far more fuel and had been specifically designed for the route. Not surprisingly, fuel on *Sirius* ran low, and the captain ordered that they burn whatever they could, including cabin furniture, spare spars and her extra mast, as well as five drums of resin; true to her home port, she persevered, day after tense, hard steaming day. It was the 23rd April when she arrived in New York with less than a day's supply of fuel on board and *Great Western* some miles behind. Nearly neglected by history, *Sirius* was the first vessel to cross the Atlantic under steam power alone.

Great Western had made a faster passage. With half her internal space dedicated to engines, boilers and fuel, she arrived later the same day and with two hundred tons of coal to spare. She had crossed the Atlantic in fifteen days and became the first steam ship to make regular trans-Atlantic crossings. Her name is remembered by nautical historians, while that of *Sirius* has been relegated to a footnote which is a shame, but perhaps proves that brash marketing catches the public eye while dogged perseverance achieves success.

Chapter Thirteen

The German Wars

The Firth of Forth has many claims to history, but not all are welcome. One unwanted accolade is being the scene of the first Royal Navy ship to be sunk by a German U-boat and the first ship to be sunk by a locomotive torpedo. It was the 5th September 1914, and the First World War was only a few weeks old. The Army was already in France, winning glory with its sacrifice of blood, but the Navy had not yet fully realised what modern sea warfare was all about.

The First World War

Deep within the British psyche was the desire for victory in the old Nelsonian style, the thunder of rolling broadsides as fleets of huge battleships pounded each other, ending with the ragged cheer of a boarding party as brave bluejackets carried the White Ensign to yet another tremendous victory against Johnny Foreigner. However, the world had changed, and such events were of the past.

At the outbreak of war, the 8th Destroyer Flotilla was based in Rosyth. The Navy designated it as a Patrol Flotilla, with one scout cruiser, HMS *Pathfinder*, thirteen destroyers and eleven torpedo boats. On the 5th September 1914, the flotilla was patrolling the Firth of Forth, escorting HMS *Pathfinder*. The Forth was a major player in the infant war, with Rosyth home to Admiral Beattie's battle cruisers, the fast moving hit men of the Royal Navy.

Pathfinder was ten years old, not ancient by warship standards, but in that era of rapidly advancing technology, she was already outdated. Even worse, she should have had a top speed of around twenty-five knots, but her commander, Captain Francis Martin-Peake had put to sea with little fuel, so she steamed at only five knots, walking pace. There seems to have been a navy-wide shortage of fuel at that time. At around noon *Pathfinder*'s destroyer escort returned to port but the cruiser limped on, splendid to look at with her two tall funnels and layers of guns, but slow and much more vulnerable than she realised to the furtive forces of the Imperial German Navy. She sailed outside the Forth, past Dunbar and down to St Abbs Head.

The Royal Navy was not the only force out that day. The German Navy had sent three submarines to penetrate Scotland's defences, and they waited in the lee of the Isle of May, much as Bull had done over four centuries previously, and with much the same objective. One of the submarines was U-21, commanded by twenty-nine-year-old Otto Hersing from Alsace. He had sailed as far as the Forth Bridge until sharp-eyed gunners at the Carlingnose battery at North Queensferry noticed his periscope and opened fire with his six-inch guns. Hersing withdrew at high speed, daunted but not discouraged. With the Forth so busy he knew there would be a target if he waited.

When Hersing saw *Pathfinder* crawling along at five knots, he must have thought it was submariner's Christmas: a slow moving large warship with no escort. It was a calm day; he was under no threat; the British vessel did not suspect a thing: perfect killing conditions for a submarine. He watched the cruiser through the periscope, manoeuvred his submarine to within 2000 yards of *Pathfinder*, lined her up in his sights and loosed his torpedo. It was a quarter to four in the afternoon.

When an alert lookout on *Pathfinder* saw the wake of the torpedo and sounded the alarm, the officer of the watch gave orders to reverse one engine and drive the other at full ahead, trying to turn the cruiser toward the torpedo so it would pass alongside and miss. But there was not enough time. Hersing had all the advantages, and his torpedo struck below the bridge, exploding on impact and causing a second explosion that tore *Pathfinder* in two. Her stern and bow both thrust upward as her back broke; there was no time to launch the lifeboats and she sunk within two terrible minutes, One survivor, Lieutenant Edward Sonnenschein, reported that the captain ordered: 'Jump you, devils, jump!' But there was no time even for that. *Pathfinder* took an estimated two hundred and sixty-one men with her. Other accounts speak of two hundred and fifty-two dead. Despite remaining with the ship, Captain Martin-Peake was one of the eighteen or so survivors.

Naturally, the explosion and resulting tall column of smoke attracted attention and the fishing boats from nearby Eyemouth as well as the St Abbs lifeboat raced to help. The scenes they encountered must have remained with them for years. The whole area where *Pathfinder* sunk was a mess of fuel oil and scattered with wreckage and pieces of human bodies. Two destroyers, HMS *Stag* and HMS *Express*, arrived slightly later after the fishing boats had rescued all who were alive. At first, nobody was sure what had happened, with speculation that it was a submarine, a German trawler disguised as a British vessel or that *Pathfinder* had struck a mine. Other accounts stated that *Pathfinder* had sunk one of two submarines that attacked her, or that the Royal Navy had later sunk the attacker. None were correct: the stark truth was that a German submarine had sunk a British warship in British waters and had escaped untouched. The reality of modern warfare had come to the Forth, and the Royal Navy knew that their domination of the seas, unchallenged for a century, faced its biggest threat.

Hersing continued his career to become one of the Kaiser's most renowned submarine commanders. He sank another three warships and became known as *Zerstorer der Schlachschiffe*, Destroyer of Battleships. He also sent over twenty unarmed merchant ships to the bottom. He may have been thought glorious, but his victims knew the reality. War is mass murder camouflaged by gaudy national flags and lauded by erudite politicians who dream up chimaeras to create real obscenity.

In 1903, eleven years before the start of the First World War, the Admiralty had selected Rosyth as a base. It was strategically situated opposite the German naval bases and could be defended from sea attack. Admiral Beatty was not keen on Rosyth, however, thinking the Forth was prone to fog, and the fleet could be trapped if the enemy collapsed the Forth Bridge. Rosyth had space for around twenty major surface craft, all part of the Royal Navy's Grand Fleet, which was also based at Invergordon and Scapa Floe. It was not until the German Admiral Hipper bombarded Scarborough, Hartlepool and Whitby in December 1914 that Admiral Beatty and his battle cruisers were sent to Rosyth to intercept any further German raids.

Britain's naval bases traditionally faced south and west to face Spain and France. By 1903 the Admiralty had realised that Germany was an up-and-coming danger and requested permission to create a base at Rosyth. From here the Grand Fleet could threaten the German base in Heligoland and menace the Skagerrak, the key to the Baltic. Rosyth was also defensible, yet despite the initial optimism, by 1914 Rosyth was considered too small for the Grand Fleet, which was based instead at Scapa Flow in Orkney. Admiral Beatty's battle cruisers, however, dropped anchor in Rosyth in December 1914 along with their escorting cruisers and destroyers. Such a fleet must have been a tremendous sight for the locals.

Beatty led the battle cruisers to engage the Germans at Dogger Bank and Jutland, with the latter an encounter of mixed fortunes. Although the Royal Navy endured the greater losses, the Germans had withdrawn at great speed, to hide behind a screen of mines and submarines until with the close of the war their fleet ignominiously surrendered with hardly another shot fired. The Royal Navy escorted the German High Seas Fleet to Scottish anchorages, and the people of the Forth had the opportunity of witnessing at first hand the enemy battle fleet that had been the bogey men of their nightmares. Eventually, the Royal Navy escorted the High Seas fleet to Scapa where its subsequent scuttling was an act that proved defiance in defeat but did not challenge the victory of the Royal Navy.

Ports such as Port Edgar also came under Admiralty control, with smaller vessels hoisting the bold White Ensign in defence of freedom. In addition, the Admiralty commandeered many fishing vessels for escort and mine-sweeping duties, while men considered too old or too young for active service chanced mines as they tried to feed the nation. Long before the end of the First World War, the Navy had a presence in every major harbour in the Forth. There were three lines of anti-submarine defences stretching across the firth, with a gap (the Fidra Gap) for access and egress, patrolled by the Navy. Burntisland harbour was heavily involved in the building work, while the Navy utilised Grangemouth and Bo'ness for fuel transportation.

Any German vessel entering the Forth would have a hot reception, with gun batteries on the Fife and Lothian shores of the Forth Bridge as well as elsewhere on the coasts including Kinghorn and Leith. There were some defences pointing inland as well, not for fear of a sudden rising of the good folk of Fife or Lothian but in case the Germans managed to land a raiding party to take the defences from the rear. Many of the islands of the Forth were also fortified, with Inchkeith having a formidable array of heavy artillery.

Although before the war attention had focussed on the race to build more and better dreadnoughts as the heavy battleships were known, the loss of *Pathfinder* and other vessels highlighted the danger of submarines. It was obvious that the Navy had to develop new skills to cope with this menace, as well as that of mines. Submarines were the new privateers; a threat that could slip beneath the guns and eyes of the Royal Navy. Offensive weapons such as depth charges were all very well but were pointless unless the Navy knew where the enemy submarines were. Accordingly, the Navy experimented with the new submarine detecting device called hydrophones at Granton and Aberdour. They established seven hydrophone stations around the Forth, linked to minefields with the idea that if an enemy submarine penetrated to the minefield, the mines could be detonated.

Local merchant seamen paid a terrible price in the four years of war, with nearly half the ships from Leith falling victim to torpedo or mine. There was a shocking tragedy in January 1918 when a flotilla of Forth based British submarines sailed out of the firth and literally ran into British surface vessels. The ships were sailing to take part in a major naval exercise but instead became involved in a series of collisions that sunk two submarines and damaged four more, as well as a surface vessel. Over a hundred men died in this episode, grimly known as the Battle of May Island.

These submarines were 'K' class – called Kalamity class by the Navy on account of the number of accidents. These were steam powered vessels that lost six of the eighteen launched through accidents, and none by enemy action. They proved equally futile in the offence, with their only recorded victory an action when a K class rammed a U-boat after her torpedo malfunctioned.

As well as the remains of the gun emplacements on the islands and various sites on the coast, there are more poignant memorials to these sad days. The Fisheries Museum in Anstruther holds a sobering list of fishermen who died at sea in that war, a reminder of the terrible sacrifice of civilians as well as soldiers. Leith also has the Merchant Navy Memorial at Tower Place, which commemorates the Scottish merchant seamen who died at sea to make the world a safer and better place.

Overall the Forth played a vital part in the Kaiser's War, providing men, ships and basis for the eventual victory, but, in common with all other areas of the country, there was terrible sacrifice regarding men killed and injured. But the First World War was not the war to end wars, as was hoped. Slightly more than twenty years later Britain was again fighting the same enemy, and once again the Forth was in the front line.

Air raid over the Forth

At first, the people along the Forth would enjoy the sight of the aircraft they saw flying in from the East. They would admire the excellent formation, point fingers and speak about exercises and the Royal Air Force. Some young boys would wave as the aircraft swooped lower down, and the airmen waved back.

The passengers on the Edinburgh to Aberdeen train that was crossing the Forth Bridge were less impressed when the aircraft came lower and began to drop bombs. It was the 16th October 1939, the Second World War was six weeks old, and the Germans had launched another attack on Great Britain. Only two days previously U-47 had penetrated the defences of the naval base at Scapa Flow and torpedoed the battleship Royal Oak, with the loss of 833 men. The attack forced the Home Fleet to shift base to Loch Ewe on the west coast. Now the Luftwaffe was once more on the offensive, and Scotland was again the target.

The Luftwaffe was not flying in the blind hope that they might find a ship to bomb. They had been searching for the Royal Navy and in particular HMS *Hood*, the pride of the fleet, ever since the 26th September when a Luftwaffe patrol spotted her in the North Sea. The Germans had tried to trace her, and on the morning of the 16th October a patrol had flown over the Forth and spotted *Hood* far below. The news excited the German high command, and they ordered Number One Squadron Kampfgeschwader 30 (Eagle Wing) to go and sink the British ship. They took off at noon from Westerland on the island of Sylt off the German west coast.

Helmuth Pohle commanded this squadron of Junkers 88s and Heinkels. At that time the Junkers 88 was the fastest bomber in the world, with its two engines giving a top speed of 270 miles per hour: the Germans had sent their best.

So despite the concerns of the passengers on the train, the Luftwaffe was not after the Forth Bridge. Instead, they were targeting the units of the Royal Navy that were in the Forth under and around the bridge, with Rosyth again a vital naval base. Twelve German aircraft had set out, but only nine made the hour and a quarter crossing, so three had dropped out on the way. Nine was a tiny figure compared to the raids later in the war, but deadly enough and very daring as they operated at the extreme limit of their range. They had no margin for error. They were too late for *Hood*, who was already safe in Rosyth, but there were other Royal Navy ships to attack including the cruisers *Edinburgh* and *Southampton* and the destroyer *Mohawk*.

The Luftwaffe was lucky in some ways. The British anti-aircraft batteries of the 94th City of Edinburgh Anti-Aircraft Regiment sat on Portobello Power Station and the Ramsay Technical College in Edinburgh. Even if they had been ready, they were not particularly formidable in that period of Phoney War and Dad's Army technology. The regiment boasted one three-inch gun of First World War vintage and a handful of creaking Lewis guns, but even these might have been useful if they had known the Germans were approaching. However, the early warning system proved useless, and the anti-aircraft gunners were caught completely at a loss. The three-inch gun was in the middle of an exercise when the Germans appeared and was loaded with dummy ammunition. Before they could fire, the gunners had to unload and reload, all of which took valuable time as the aircraft were roaring to their target.

In the meantime, the passengers on the Aberdeen train had a grandstand view of the whole battle. As the first German planes flew parallel to the bridge, and then the bombs fell; the train stopped on the bridge and the workers, the painters and riggers, hastily climbed down from the scaffolding on the orange-red steel girders and looked for whatever cover they could find. Much to the relief of the passengers, the train started again and steamed out of the danger area.

As the train left the bridge, the Navy's anti-aircraft gunners opened defensive fire. HMS *Mohawk*'s pom-pom did not fire, but her machine guns let rip as a German plane dived on her, dropped two bombs from about 600 feet and machine gunned the ship's bridge and upper works. Both bombs exploded on contact with the sea, just as the crew was mustering for action stations. The casualties included the first officer, who was killed, and Captain Jolly, who was wounded in the stomach. The men around the searchlight, the mooring party on the foc'sle and the men at the after control position also suffered.

Pohl's aircraft released a thousand pound bomb that crashed through three decks of HMS *Southampton* but luckily failed to explode; only the admiral's barge was destroyed. *Mohawk*'s machine gun had also been accurate, destroying the cockpit escape hatch of the leading Junkers. Despite his wound, Jolly retained command of *Mohawk*.

'Leave me' Londoner Jolly ordered in the best tradition of the Navy, 'go and look after the others.' He brought *Mohawk* into Rosyth but later died of his wounds. He was awarded the George Cross.

The German fliers were brave men. They bombed with accuracy and did considerable damage to three ships: the cruisers *Southampton* and *Edinburgh* as well as *Mohawk*. In total sixteen seamen were killed, mostly on *Mohawk*, with forty-four wounded, but the government did not immediately release these figures to the public.

It was not all one-way action as 602 City of Glasgow squadron scrambled from Drem in East Lothian and 603, City of Edinburgh squadron from Turnhouse. Neither were regular squadrons but auxiliaries composed of eager weekend airmen desperate to do their stuff and prove their mettle against the feared Luftwaffe. They met the Germans in the air, and the Firth of Forth became the scene of the first dogfight of the war between British and German aircraft. The motto of 603 Squadron was Gin ye Daur – if you dare - and they pitted their skill against the Luftwaffe over the old Scottish Sea.

Flying Spitfires, they tore into the German aircraft, with Flight Lieutenant Pat Gifford leading the first wave of Spitfires and personally finishing off the first Ju-88. In the meantime, 602 Squadron attacked Pohle's Junkers. Flight Lieutenant George Pinkerton and Archie McKellar brought him down off Crail. The British won the aerial fight, shooting down German aircraft for no losses; these were the first Germans shot down over British territory in the war and the first German aircraft downed by a Spitfire, the aircraft that was to have such a huge impact during the Battle of Britain the following year.

The last of the German aircraft was shot down over May Island and a trawler rescued two survivors from the Forth. The German airmen were fairly cheerful and seemed very confident that they would defeat Britain within a few weeks. Pohl had proved his bravery; now he showed his arrogance by saying he was a friend of Goering's and demanding the British fly him back to Germany in a Red Cross plane as he was wounded. Not surprisingly, the British refused this request. Other captured German airmen spoke of the 'peasants' who had waved to them, which was perhaps not the most diplomatic way to talk about Scottish civilians during a war. The German airmen who died were buried with full military honours in Portobello, with a piper playing a lament.

After opening the ball above the Forth, 603 Squadron took part in the Battle of Britain in 1940, the sweeps over France in 1941, and the defence of Malta in 1942. HMS *Cossack* was a Tribal class destroyer; she later took part in North Sea convoys and was present at the Battle of Calabria in July 1940 and the Battle of Cape Matapan in March 1941. She was torpedoed and sunk by an Italian destroyer off Tunisia in April 1941. HMS *Edinburgh*, a Town class cruiser, took part in the Norway campaign in 1940, was on convoy escort duty to Africa, Russia and the Mediterranean, hunted for the German battle cruiser, *Scharnhorst*. In 1942, returning from Russia with a cargo of gold bullion, she was sunk by a combination of German torpedoes and a flotilla of German destroyers. It was not until the 1980s that most of the gold was recovered. HMS *Southampton* also hunted for *Scharnhorst*, served in the Norwegian campaign and was involved in anti-invasion duties off southern England. She also took part in the Red Sea campaign against Italian East Africa and was sunk off Malta in January 1941.

During this war as in every war, the Forth sent out its finest. One of the best was Captain Eric Melrose Brown (1919 – 2016), known as 'Winkle' who was a Leith man and was in Germany at the outbreak of the Second World War. Temporarily imprisoned by the SS, he was released and on his return to the UK joined the Fleet Air Arm, alternating between operational flying and test flights. He was highly decorated, flew 487 different types of aircraft and was the first pilot to land a twin-engined aircraft and a jet on an aircraft carrier. A fluent German speaker, he also interrogated various high ranking German officials after the war, including Josef Kramer and Irma Grese, who ran the Belsen concentration camp as well as Hermann Goering. He survived the war and did not die until 2016.

The war continues

The attack on HMS *Hood* was only the first stages of the Forth's war. U- 21 laid magnetic mines in the Firth and on the 21st November 1939 HMS *Belfast*, a new light cruiser, was severely damaged. Belfast built, she was launched on St Patrick's Day, took part in Arctic convoys, in the destruction of *Scharnhorst*, in the Normandy landings and against the Japanese as well as serving in the Korean War. She survived all that and now floats on the Thames as a museum. As in the earlier French wars, the Forth was the base for convoys, either down the East Coast – E-boat Alley- to London or around Scotland to the Clyde. There were losses on both sides as the Germans mounted air, submarine and surface attacks on the British vessels and the British struck back. There was a notable success in March 1945 when the newly commissioned South African frigate HMSAS *Natal* sunk U-714 off the firth. Another came next month off St Abbs Head when U-1274 attacked Convoy FS 1784 and sunk a merchant ship before the escorting destroyer HMS *Viceroy* accounted for her. The submarine sunk with all the men on board. The Forth also witnessed the last Allied ships sunk by a German submarine in that war, when U-2336 torpedoed the Canadian *Avondale Park* and Norwegian *Sneland 1* just inside the Isle of May in April 1945.

The Navy's here

There were many other memorable incidents during that war. On the 9th November 1939, the Luftwaffe made an ineffectual but worrying attack on Leith, a prelude to the far more destructive raids that were to create such havoc in Glasgow and Clydebank, London and Coventry in the years to come.

On the 17th February 1940, the Tribal class destroyer HMS *Cossack* sailed into Leith with 299 British merchant seamen, rescued from German captivity. The seamen had been the crews of British ships captured by the German pocket battleship *Graf Spee*. The Germans transferred their British prisoners to a tanker named *Altmark*, which sailed through neutral Norwegian waters. After futile searches by the Royal Norwegian Navy, the RAF spotted *Altmark*, which hurried to hide in Jossingfjord, but the Royal Navy pursued her. On the 16th February 1940, Captain Philip Vian of HMS *Cossack* sent a boarding party with bayoneted rifles and the last recorded uses of cutlasses in Royal Navy history. The German crew fought back, wounding one British seaman in return for a reported four German dead and wounded.

Once German resistance was overcome, the Navy searched the ship for any British prisoners. In response to an enquiry, the merchant seamen shouted:

'We are all British here!'

'Well, the Navy's here!' Replied the men from HMS *Cossack,* and the prisoners set up a huge cheer. *Cossack* was later involved in the Second Battle of Narvik, and the destruction of the German pocket battleship *Bismarck* before the Germans torpedoed her in the Atlantic in October 1941.

Crail was also involved in offensive operations. During the First World War, a naval air station was built just outside the town, which became HMS *Jackdaw* as Hitler's War loomed on the horizon. Although it was mainly a torpedo training school, aircraft took off from here during the campaign to sink *Tirpitz* in 1944. On a more routine day, Jackdaw's aircraft provided air cover for convoys from Leith and Methil.

Slightly further north, and perhaps more connected with the Tay than the Forth, the airfield at Leuchars has a claim to fame, for on the 4th September 1939 a Leuchars based Lockheed Hudson of 224 Squadron became the first British aircraft to battle the fearsome Luftwaffe. The Hudson was an American built aircraft that became a mainstay of Coastal Command. It was patrolling over the North Sea and encountered a Dornier DO 18, and both aircraft exchanged shots without much damage to either side. It was a Hudson flying from Leuchars which first saw *Altmark* later that year. Aircraft from Leuchars were also involved in patrols against German shipping and submarines in the North Sea.

Naturally, with the fear of invasion, the coasts of the Forth were fortified and armed. For example, quiet Earlsferry had naval guns on the cliffs, observation points near the lighthouse and a machine gun nest at the west end of the beach, ready to sweep Hitler's army away if they invaded from Norway, as was feared. The major islands also carried fairly formidable artillery in case the Germans penetrated to the Forth.

All around the Scottish coast, including the Forth, enormous blocks of concrete were situated to prevent, or at least slow down, any hostile tanks or motorised transport that may try to invade. While negotiating these obstacles, the vehicles would be vulnerable to artillery fire. There was the black out, of course, and while local men joined the British Army, Polish soldiers were based and trained in Fife, ready to defend Scotland.

But the Forth did not just act as a battleground and a base for offensive and defensive operations. It was also a busy industrial hub, with the shipyards flat out in ship building and ship repair work. The Burntisland Shipbuilding Company built cargo ships, including two a year for the Carlton Line of Newcastle, colliers for the Gas Light and Coke Company, frigates and a significant number of merchant aircraft carriers for the Royal Navy, War Standard vessels for the government and other craft. Leith also built and repaired ships, with Henry Robb building forty-two vessels for the Royal Navy and fourteen merchant vessels, as well as repairing around three thousand other craft, so the yard was always busy. Robb built the workhorses for convoy protection: corvettes and frigates, plus minesweepers, trawlers and tugs. Leith also built some of the pierheads and sixteen pontoons for the Mulberry harbours that made a significant difference at the D-Day landings.

Overall, the Forth was on the front line in the Second World War, as a mustering point for convoys and as a target for the enemy. Rosyth, of course, was a major naval base, and there were ship repair and shipbuilding yards around the firth. Most important of all were the people, defiant, stubborn; the descendants of the men and women who had fended off Vikings, English and French and now equally ready to cross sword with Hitler's Germany. Undemonstrative and sometimes dour but never willing to bend before tyranny, the people of the Forth played their part and more in bringing Hitler to heel.

Chapter Fourteen

Crossing the Forth

Leith is the port of Edinburgh, able to rival her inland rival and partner in age and ancestry. Since the latter decade of the twentieth century, Leith has undergone something of a renaissance, with grand new offices and upmarket wine bars, but it was not always so. At one time Leith was a town in decay, slightly hazardous on a winter's evening and with pubs that it might be better to avoid. One such was the Willie Muir, on West Granton Road, just outside Leith and a hostelry that was famous, or infamous, throughout the area, with rumours of a couple of murders there in the 1970s. The stories were probably apocryphal, but even so, it was not a place that visitors would have been advised to enter. Others, who used the establishment as locals, thought it friendly.

Willie Muir; it's a common enough name that probably many thousands of decent Scots have used, but somehow the very ordinariness seems to place it firmly in the Forth.

Willie Muir

The Willie Muir public house took its name from *William Muir*, which was the longest lasting and arguably the most famous ferry to operate on the Forth. The ferry Willie Muir made some 80,000 crossings of the Forth. From 1876 to 1837 she crossed from Granton to Burntisland, with the occasional break for incidentals, like fighting the First World War. She was a Forth legend whose withdrawal from service in March 1937 spawned genuine grief and some interesting poetry. As the local songwriter Nan MacDonald wrote:

We hailed her as a well-known friend.
The Burntisland boat

We deemed her quite the finest type
Of ferry-craft afloat.

Built at Kinghorn, powered by engines from Leith and named after the chairman of the North British Railway, *William Muir* was a Forth boat, through and through, and experienced her quota of incidents. She was involved in the infamous 1879 Tay Bridge disaster, with the passengers who died when that bridge collapsed having sailed on her across the Forth. In 1889 she had what was possibly her most memorable cargo. *William Muir* was one of two ferries that transported Lord George Sanger's Circus of fifty caravans and around five hundred animals in an overnight succession of crossings. The seamen were a bit innovative in their loading techniques, using the elephant, named Jumbo, to push the largest caravans on board.

She continued her crossings for the first three years of the First World War, despite the difficulties of manoeuvring through minesweepers, patrol boats and other incomers to her patch of water. In 1917 the Granton to Burntisland ferry was suspended, and the Admiralty snatched *William Muir*, transformed her into a minesweeper and sent her south to Sheerness, with her own master Captain Clark, remaining in charge. Returning after the war *William Muir* resumed her old job until, after some 800,000 miles, her owners decided to retire her. This famous old vessel, however, was only one of a host of forgotten or near forgotten ferry boats and cruise ships that plied the Forth, many of which deserve to be better remembered.

The ferries, of course, date back at least to the time of Queen Margaret, that enigmatic, saintly person who brought Scotland into the mainstream of western European religion while simultaneously eroding much of the Celtic Christianity that had served so well. It was in 1164 that the name 'Queens Port of Ferry' was first recorded and twenty years later a Papal Bull of Lucius III confirmed the name Queensferry. At that time the Church controlled the ferry, but by 1275 it was back in the care of laymen. Or rather lay people, for two of the eight people who hauled at the oars were women. Mediaeval women, of course, also physically carried people across many Scottish fords. In 1474, after much fluctuation, the ferry fares were set at a fixed rate; passengers paid one Scots penny for the passage, while horses were twopence. Sometimes the ferrymen had strange adventures, such as the sighting of a whale in 1843. After a frantic row to the shore to pick up harpoons, the ferrymen fought an hour-long battle and succeeded in killing the unfortunate animal.

The Queensferry was to be the longest lasting ferry of all, for it survived until 1964 and the advent of the Forth Road Bridge. After opening the Bridge, the Queen and Prince Philip boarded the last ferry, appropriately named *Queen Margaret*, and sailed to Hawes Pier.

There were many other ferries. Like the Kinghorn boat – notice the incredibly unromantic title that Forth dwellers apply to their environment?

Kinghorn was the Fife port for the ferry, otherwise called the Broad Ferry. In the early days, the departure point was regulated solely by the weather, and passengers might have arrived at Kinghorn, only to find that the ferry left from Pettycur, or vice versa. Customer care was not a particularly high priority for the four bitter tongued ferrymen who manned this boat, for they often laughed as their passengers swayed up the narrow gangplank that was the only means of boarding the ferry. Not surprisingly this procedure was known as 'walking the plank'. At one time the ferrymen were a law to themselves, but an Act of 1425 set formal rules that they should 'be ready to serve all men…Raise, nor take more fares of our sovereign lieges for man, horse or goods but as much as in statute.' For passengers who survived the ordeal of the plank, there was a crossing that could take up to six hours, before a thankful arrival at Leith or Newhaven. There was a royal connection here too, for in 1589 one of the vicious Forth squalls capsized the ferry boat and drowned forty passengers, including Lady Jean Kennedy of the court of King James. A similar storm befell a vessel of King Charles I when he sailed from Burntisland on Wednesday, July 1633. The vessel was the Dutch designed, Blessing of Burntisland, built of timber from the Baltic, as was not uncommon in east Scotland. The King had been on his coronation tour of Scotland and was returning to England. In a typical king-like manner, he carried a large amount of treasure with him, estimated at £100,000 worth of plate, kitchen equipment, gold and silver plate for dining and the royal wardrobe. That was a massive amount of wealth at the time and would be worth millions at current values; all ended up under the waves of the Forth. The clothes from the wardrobe will be long rotted, but the rest would be invaluable if some enterprising person could recover it. Only two men are supposed to have survived the typical Forth squall.

Naturally, there have been attempts to locate the treasure ferry (only Scotland could have a treasure ferry and not a treasure galleon) but without success so far.

Overlooking Pettycur is a green hill, which had the local name of Crying-oot Hill. According to legend, the name derived from the days of the ferry, when a man stood on the hill and cried out when the ferry was coming. The other passengers, who were snugly ensconced in the local pubs, would hurry to the pier. However, it is also possible that the hill was a lookout spot for fishermen searching for shoals of herring. Either possibility has its interest.

Dangerous voyage

Sometimes, people had adventures crossing the Forth.

At half past seven in the morning of the 20th February 1816 the ferry boat *Providence* set off from Kinghorn to Leith. There was a full westerly gale howling down the firth, but the superintendent believed it safe, despite the coast of Lothian being nearly invisible due to the lashing rain and spindrift kicked up by the wind.

When she was about half way across it was evident that *Providence* could not make Leith in safety, so the master steered for Fisherrow instead. However, the tide had turned during the difficult crossing, so he decided to head for the lee of Inchkeith to ride out the storm. As they tacked toward the island, *Providence*'s main boom snapped in half, and she bucked and tossed to the tune of the storm. The master regained control, avoided the rocks of Inchkeith, lashed the helm and allowed *Providence* to run before the wind. He hoped to make the open sea where there were no islets or dangerous rocks.

By that time in the late afternoon, the sea was in a frenzy, with waves higher than *Providence*'s mast and the passengers reconciled to die out here, away from their families and friends. Once he thought they were safe, the master dropped both fore and aft anchors to ensure they were not pushed too so far out to sea that they could not get back.

The people on board *Providence* were in a bad way. They had boarded her for the short crossing from Fife to the Lothian coast, so had neither foul weather gear, food nor water; they had not prepared for a prolonged stay at sea. The anchors did not help, and *Providence* still drifted, so the master decided to try for Elie, on the Fife coast. He hoisted one anchor with difficulty, but the cable of the other snapped and he had to abandon it to the sea.

The master had never left the helm, and now he ordered the crew to redouble their previous already strenuous efforts. They followed with a will, and *Providence* crept into Elie harbour, still in Fife, tired, wet and hungry but without a single casualty.

Riotous students

Students can have a reputation for somewhat unruly behaviour from time to time, and that was as true in the nineteenth century as it is today. On Saturday 19th June 1853 some students of botany were crossing between Burntisland and Granton on the ferry. The whole party numbered around a hundred, under the rather loose control of a flustered Professor Balfour. They had spent the day in Fife, and a number had found the local hostelries very welcoming. When they boarded the ferry, some of them ran to the paddle box and climbed on top, which annoyed the master, who ordered them off. The students objected and argued back, but the Master of a Forth ferry was equal to a gaggle of teenagers and soon had the paddle boxes cleared of the unwanted intruders.

It was only a few moments later, with the ferry under way and the water churning frothy white around the paddles, that one of the crew mistook Professor Balfour for one of the students and gave him a mouthful of less than respectful nautical language. Some of the students heard and reacted by attacking the seaman. The sailor fought back but he was outnumbered, and the students gave him a hard time. Naturally, other seamen joined in to support their colleague, but there were far more students than crewmen, and they were getting the better of things before the master intervened.

Rather than rush in with flying fists and a belaying pin, he gave an order to attach the hose to the boiler so that it could issue a powerful jet of hot water. As soon as the other passengers realised what might happen, they rushed to clear the deck and escape to the cabin, with the women prominent. There was a mad melee of rustling skirts, prodding parasols, sharp elbows and sharper tongues as the women out-muscled the men in their scramble for sanctuary.

In the confusion, somebody broke the cabin windows, but whether by students or panicking passengers nobody knew. The master watched the chaos, took hold of the hot-water hose and stationed himself on the gangway, standing between the two paddle boxes. The students saw the weakness of the master's position and formed a plan to regain control of the ferry. They made a mad rush forward, and while some diverted the captain's attention, others drew knives and slashed at the hose so that boiling water spurted out in half a dozen places and pooled on the deck. Despite the crowd in the cabin, many of the passengers were still on deck, and the boiling water splashed them, causing some unpleasant scalds.

As the passengers yelled at their burns, the students cut the hose into sections and threw it into the Forth. All the while the ferry chugged on at top speed, with the paddles churning the water into as much a frenzy as the people on board. As soon as the boat reached Granton, the students rushed to the shore, with the captain and crew too glad to see them go that there was no attempt to hold any for the police.

Is there a moral to this story? Perhaps not; only a reminder that youth has always been high spirited.

Railways and pilgrims

Interestingly, the arrival of the railways did not signal the immediate demise of the Forth ferries; there was a period of cooperation between land and water when, in 1850, the Burntisland ferry became the world's first rail ferry. Thomas Bouch, who later was to design the ill-starred first Tay Bridge, designed this 'floating railway'. Named *Leviathan*, it operated on a roll on, roll off system but must have been quite alarming to the early users.

There were many reasons to use a Forth ferry. In mediaeval times pilgrimages were a popular way to see the world, and incidentally, help save one's soul, and in Scotland, there were few better places to visit than St Andrews. North Berwick was the ferry port on the south shore of the Forth, but it would be a dangerous pull around Fife Ness, so the ferries sailed to Crail or Earlsferry. As the mouth of the Forth has more than its share of bracing weather, even this journey must have been quite an experience. The name Earlsferry is nearly as interesting as the history of the village.

Earlsferry supposedly gained its name from MacDuff, Thane of Fife, who slipped through an underground passage from MacDuff Castle to the Well Cave near East Wemyss when Macbeth was happily massacring his family. MacDuff followed the cave to what is now Earlsferry, from where he crossed to North Berwick and fled to England. There is a tradition that Malcolm Canmore repaid MacDuff's loyalty by decreeing that anyone suspected of crime must not be pursued from Earlsferry till he was halfway across the Forth. Nearby to Earlsferry is the ruin of an eleventh-century chapel that pilgrims used when they travelled to St Andrews, which again has a MacDuff connection, for the Earl built the chapel to show his gratitude for the ferrymen saving his life. The chapel was well patronised, for an estimated 10,000 pilgrims crossed here annually.

Sailing for fun

The Forth was not all about work; there were pleasure boats here as well. On Tuesday 22nd July 1812, despite the ongoing war with France, there was a sailing match between two pleasure boats, *Jessey* and *Wellington*. They left the pier head at Leith at one in the afternoon, sailed past Inchkeith and round the buoy at Craigwaugh and back, which was a circuit of seventeen miles. Although there were just two boats in the race, a huge crowd gathered to watch, as Captain Skene in *Jessey* beat the favourite. Such events broke the monotony of hard-working lives, gave expression to natural competitiveness, allowed pick-pockets access to scores of unwary customers, and taught caution to the gullible, so everybody gained.

Sometimes people sailing in the Forth met with more adventure than they wanted. In September 1825 James Greig of Kirkcaldy took his boat out of Burntisland harbour for a small cruise in the firth. He had no object other than enjoying the day and catching a few fish. When he saw some seagulls floating on the surface of the sea he guessed there might be a shoal of mackerel beneath, so readied his fishing lines.

Greig steered for the centre of the patch of birds and stared in astonishment when his boat jarred to a halt. He looked over the side and realised he had run the keel onto the body of a large whale. Unable to move, Greig could only stare as a huge fin rose beside him, followed by the rest of the whale; eight, ten, maybe fourteen feet out of the water. He watched, unable to do anything as this underwater giant took control of his boat. Eventually, the whale rolled over and swam away. Greig reckoned that it was at least fifty feet long, but it did him no harm; strange things could happen in the Firth of Forth.

From the early nineteenth century, a new phenomenon burst into the Forth in the shape of pleasure cruises. While before the Firth had been a place to cross as quickly as possible or a place to make a living by trade or fishing, now it could be seen as a place to enjoy. What opened the water up to more people was the advent of steam power, as pioneered by Scottish engineers such as Henry Bell in his famous Comet.

Many of these early pioneers were outrageous; sober accountants in a frock coat and top hat would not speculate on anything as bizarre as a new method of propulsion. But given the prosaic nature of the Forth, perhaps it is not surprising that the wildest of them was not a local, but came from the North East of England. In the late 1860s, George Jamieson was joint owner and master of the steam tug *Garibaldi* that took passengers from Leith to Aberdour to enjoy the Silver Sands. At this period Britain was deeply divided over temperance, with many 'respectable' people denouncing drink as an evil thing. Public houses were no longer allowed to open on the Sabbath, but Jamieson spotted a loophole in the law. The Licensing Acts did not bar drinking at sea, as *Garibaldi* became so famous for supplying alcohol that she earned the sobriquet of the 'floating shebeen'.

In August 1868 she became the sinking shebeen when *Lord Aberdour*, a steamer owned by the rival firm of John Galloway, accidentally rammed her off Newhaven. It is hardly surprising that there was much acrimony between the two companies. Refloated, *Garibaldi* added Saturday trips to South Queensferry to her itinerary but was again sunk, this time off North Berwick.

By 1874 Jamieson was the sole owner of *Fiery Cross*, a small but speedy tug that he painted bright red save for her tall white funnel. Between the two of them, Jamieson and *Fiery Cross* brought the Forth cruise programme alight. From 1874 *Fiery Cross* steamed from Leith to North Berwick, circled the Bass Rock and returned. She also visited Burntisland and had sightseeing trips to Inchkeith, which was then being fortified to defend the Forth. Remembering the first sinking of *Garibaldi*, in 1876 Jamieson added *Fiery Cross* to the companies that steamed to Aberdour on Sundays, in direct competition to the Galloway ferries. Jamieson used every trick he could to attract passengers onto *Fiery Cross*, which in truth was not as comfortable as the other ships on the route. The Galloway Company responded in kind, and when Jamieson believed that John Eunson, the pier porter for Lord Aberdour had called *Fiery Cross* 'an old rotten tug, not fit to carry passengers' he became extremely aggressive. After treating Eunson to a mouthful of common abuse, Jamieson offered to 'split his face' which, on a Sunday, was too much for the Leith police to dismiss. Jamieson was fined £1, with a caution of £3 to keep the peace.

Aggressive or not, Jamieson and his *Fiery Cross* was popular with the paying customers, who enjoyed the cruises around the Bass and to May Island. Perhaps it was the fact that Jamieson burned a wooden cross at her stern on the final run of the year so that the flames would reflect from the dark waters of the Forth. Perhaps it was just that *Fiery Cross* had a charisma that was more important than mere luxury, but she gained nicknames from her public. *Fiery Cross* was known as Granny's Washtub or the Aberdour Puddock, and Jamieson took her all over the Firth and beyond. She visited St Andrews and Inverkeithing, Kirkcaldy and Alloa and Stirling. Jamieson took passengers to Queensferry to view the building of the Rail Bridge and round to the Tay for a few hours in bonnie Dundee. In her way, *Fiery Cross* was a predecessor of *William Muir*, and indeed the two vessels were working at the same time.

The last cruise of *Fiery Cross* was to view the newly completed Forth Bridge, which is sad, in its way, but sentiment cannot compete with hard- hearted progress. Like *Willie Muir, Fiery Cross* is a happy memory of the Forth, but one tinged with nostalgia, for it is unlikely if any vessel sailing today will ever claim the affection held by the boats of the past.

Chapter Fifteen

The Bridges

There is something magical about the Forth Bridge. It is a symbol of unity that connects two sides of a river or two parts of a nation; it is a national icon, a poem in steel and stone, one of the major engineering successes of Victorian Scotland. It screams success and confidence and sheer brash skill. It has been called the Eighth Wonder of the World and Scotland's Eiffel Tower. It survived both the Kaiser's War and Hitler's War and is the only bridge across any of Scotland's firths that never closes, however, foul the weather. Yet it is only one of the bridges that leap across the Firth of Forth. Perhaps it is the most admired, but that is a matter of opinion.

Some bridges are merely functional, mundane structures that do their job. Bridges on motorways are like that, superbly engineered to carry vast volumes of traffic in safety, but in appearance, most are square structures that whisper utilitarianism amidst millions of tons of concrete and uninspiring, multi-lane roads lost in the overpowering entirety of the motorway system. The skills of the engineers are forgotten in the rush to travel from A to B at the fastest possible pace and with the least possible inconvenience.

The Forth Bridge is not like that. Crossing the Bridge, even viewing the Bridge, is an event, a highlight in any journey from north to south or south to north. It is symbolic of the unity of Scotland, a triumph of man over nature, for while the Forth has been both a barrier and a highway, the Bridge spans both to carry rail traffic for well over a century of rattling railways, from the splendour of steam to the speed of the present day.

In the latter part of the nineteenth century, the railway was king, and the steel lines crossed and re-crossed the country. The railway navigators, these splendidly raucous men, had worked like drunken Trojans to dig and blast cuttings, raise embankments and level ground, so the great iron rails penetrated even the most remote fastnesses of Scotland. But there were still gaps in the network. At the great Firths of Forth and Tay, the railway lines came to an abrupt halt, to recommence on the further side. In between, there were ferries, romantic and evocative perhaps, but they delayed traffic and impaired efficiency and speed. There was even the world's first roll-on-roll off railway ferry joining Burntisland and Granton, but still, travellers wanted more. Something had to be done about it, and being Victorians and therefore bustling with energy and confidence, the engineers addressed the problem.

The obvious solution was to build a bridge across the Forth. The idea of bridging the firth was not new. There had been talk of a tunnel as early as 1806, possibly sparked by the notion of the Napoleonic French tunnelling under the Channel to invade England. A suspension bridge had been considered as early as 1818, but the technical difficulties were beyond the capabilities of the engineers of the time. In the 1860s, the idea was again proposed, with a bridge built on fifty piers proposed, to be built near Blackness on the south shore to a point just west of Charlestown in Fife. That particular scheme ended, and there was a lull of a decade or so while the two rival railway companies, the North British and the Caledonian, wrangled, scrabbled for support and proved that railways provided excellent feasts for hungry lawyers.

Even the names argue for a rivalry. The Caledonian Railway has a nationalistic, Scottish twist that was augmented by its motto: *nemo me impune lacessit* (No one assails me with impunity, or wha daur meddle wi me- Scotland's motto). On the other hand, the title North British harks back to the time when Scots cringed at announcing their country's name and adopted the sobriquet of North Britain. The fact there was never a corresponding 'South Britain' speaks volumes for the nationalism of the southern nation in the political union of the United Kingdom. Scotland embraced the idea of Britain while the south remained forever England.

The Victorians were maniac about their railways. The double iron lines that crossed countries and continents represented progress, that manna worshipped by earnest, bearded men in an age when religion was tangible, and the idea of respectability often cloaked the corruption and double-dealing common to politicians and businessmen alike. The right of shareholders to pocket profit was inalienable – or so the shareholders believed – and businesses such as railways were there to make profits. There was another, more moral, reason for early Victorians obsession in railways: capital invested in the railways was better than capital invested in slave plantations in the United States. The Victorians lauded moral purity nearly as much as they loved progress and profit.

By the late 1870s, all seemed settled. The North British Railway would push the bridge across the Forth and operate the trains. A railway bridge had recently crossed the Firth of Tay, railway trains were rattling across Fife, and all that was needed was the Forth crossing to connect London to Aberdeen via York, Edinburgh and Dundee: a splendid transport link from the heart of Empire to Victoria's favoured holiday haunt. The man for the job was obviously Thomas Bouch, the English civil engineer who had designed the roll-on, roll-off train ferry that already crossed the Forth between Granton and Burntisland and, more directly, the designer of the Tay Bridge, linking Dundee and Fife. It seemed evident that he was aware of the problems of large scale bridge building and could find the solutions. Unfortunately, nobody scrutinised Bouch's other works or queried that he used cheap methods and poor quality materials to keep down costs. The spectre of shareholder profits and engineering reputation was about to cast the Reaper's grim shadow across the ethereal light of the Tay.

Work had barely started on the Forth Bridge in 1879 when it abruptly halted. On the night of 28th December that year, Bouch's splendid Tay Bridge collapsed, taking with it a train and its entire contingent of passengers. There were no survivors. Not surprisingly, confidence in Thomas Bouch's ability to design the Forth Bridge disappeared as abruptly as the Tay Bridge had and the Forth remained un-bridged for years. But passengers still expected more than a mere ferry across the Forth; they wanted a bridge. Even more than that: they wanted a bridge that would not break when a train was crossing. Sir John Fowler and Sir Benjamin Baker were selected to design this solid structure, while Sir William Arrol was appointed as the actual engineer.

In 1882 work started again, slowly and carefully. After all the early wrangling, the site chosen was from South to North Queensferry, the shortest distance across and above the old ferry route that Queen Margaret used so many hundreds of years ago. The bridge was to be built on the cantilever system, with steel girders creating a tracery across the firth above the island of Inchgarvie.

Building the Bridge

There has been a book written purely about the men who built the Forth Bridge, and quite rightly so. An international workforce came to the Forth to build the bridge. Known as 'Briggers' these men performed amazing feats as they put together the girders that formed the three double cantilevers of the Bridge. To show their standing, many of the men wore rings made from the steel of the Bridge, a fitting show of pride in their skill and workmanship.

Preliminary work began in February 1882 when engineers placed a wind gauge at Inchgarvie Castle, part way across the firth, to measure what the pressure would be when the Briggers completed the bridge. There was to be no repetition of the Tay Bridge fiasco. That same month the anticipated cost of the bridge was announced: £1,730,000, which was a huge sum for the time. In December that year, the contract for the bridge was signed and all eyes focused on the stormy, beautiful passage where the Queen's Ferry had long run. Almost immediately the builders erected huts for the workers at South Queensferry and took over Inchgarvie.

A roof was placed on the old castle and the sturdy stone walls utilised to create offices for clerks and engineers working on that section of the bridge. By the summer of 1884, there was also a canteen for the Briggers who worked and lived here. However, all this activity created a problem that had nothing to do with engineering. For generations, the Dundas family had owned Inchgarvie, and Captain Dundas of Inchgarvie objected to this takeover of his family's little island. In April 1884 Dundas demanded £25,000 for his rights to Inchgarvie. The Company refused a figure they thought was ridiculous and handed over £1,500 instead.

It took three years for the toiling Briggers to create the supports for the bridge itself. They had to sink caissons into the bed of the Forth, and men worked in these massive iron tubes, often with compressed air the only defence against encroaching water. Sometimes there were disasters and casualties among the men working beneath the surface of the Forth. Not until 1886 did work on the incredible iron bridge itself begin, with men working far above Inchgarvie and the sparkling waters of the Forth. Both these stages created various engineering problems, and both were highly dangerous to the men involved.

There were other, non-engineering problems as well. Throughout the nineteenth century, the railway navvies had a bad name from drunken violence. The name 'navvies' comes from the original labourers who built the canals or 'inland navigations' who became known as navigators, which was shortened to 'navvies' and passed on to their lineal descendants in transport creation, the men who built the railways. Now some of the Briggers seemed determined to add to the railway navvies sometimes fearsome reputation.

The camping of several hundred single men on the fringes of a small town such as South Queensferry was bound to cause some problems, particularly at weekends. By September 1883 the town council was up in arms at the amount of Sunday drinking. The blame was laid squarely at the feet of the Forth Bridge contractors, who had opened a canteen in the workers' camp. Drink and lack of recreational facilities combined to create some problems for the douce towns-folk of South Queensferry and their compatriots in southern Fife. Sundays seemed to be the day the Briggers chose to hit the beer, with, for example, one Sabbath in April 1884 seeing three drunken men fall into the sea and the police becoming involved in seven separate fights.

There were many cases of Briggers falling foul of the law. For instance, in October 1883 William Robertson attacked Constable Smith who had the unenviable task of policing the Forth Bridge Works. Robertson had fourteen days in jail to ponder the inadvisability of his actions. Such assaults were routine, such as the case in March 1887 when Bartholomew McEwan and William Mc-Parlin attacked a Glasgow man in North Queensferry and ending up with a month in jail each. There was also the husband and wife duo of Samuel and Mrs Jones who in May 1884 attacked a policeman and spent the next seven days in jail. The local publicans may have welcomed the extra custom, but there were drawbacks, such as a Brigger attacked Mr Nisbet of the Albert Hotel in North Queensferry in February 1886 or when Thomas Young attacked Mrs Wilson of the Forth Bridge Hotel in August 1887. These are only examples; there were many more such cases.

Other crimes were less open, such as the case of John Kennedy who in May 1884 went on the spree with his colleague Matthew Hunter. Kennedy could hold his drink better so when Hunter fell asleep, Kennedy robbed him of his watch and all the money he had in his pockets. The police picked Kennedy up in Dunfermline. Other thefts were equally mean, including that of a box that held the brigger's contributions to the Sick and Benefit fund. When the box disappeared from its place in North Queensferry, Inspector Arthur Webster of the county police arrested three Briggers in South Queensferry. There were other thefts, including a break in at the railway at Inverkeithing when two Briggers purloined 600 rail tickets, popped over to South Queensferry and stole a horse and cart, perhaps to carry their spoil.

Not content with theft and common assault, the Briggers added to their catalogue of crime with various other offences. For example, there was David Brown, a labourer who married Christina Ross of Dunfermline, but neglected to tell her that he was still married to a woman in Lancashire. Or there was the habit the Briggers had of jumping from moving trains just for the fun of the thing. Or the riot in Dunfermline High Street in November 1886 that saw some Briggers smash in the door of the Museum Tavern while others put the boot into Constable Hunter. Or the much more serious indecent assault on a factory girl that saw John Gorman fined a paltry one shilling in November 1886. Eighteen months later William Taylor was arrested for luring a seven-year-old girl into a railway cutting, where he sexually assaulted her, while that same year John Murdoch was jailed for the same crime on another young woman. There was also the occasional poacher, such as Alex Turnbull. The police arrested him in December 1887 for being in possession of a ferret and ten rabbit nets. He was just out of jail for a poaching offence, but a merciful sheriff let him off with a fine.

But the Briggers had other problems. Despite the crime wave, most were merely hard working men doing a dangerous job, and there was a horrendous catalogue of accidents. The contemporary press covered the building of the bridge and often carried pieces about the workers. For example, this short entry was in the Dundee Courier on the 29th November 1883: 'a boy named Thomas Harris, about 16 years of age, engaged at the jetty in connection with the Forth Bridge, has gone amissing and it is feared he has been drowned.'

Injuries were varied and frequent; with, for example, lengths of timber rolling on a foreman saw miller named Lambie and crushing both his legs. That was in December 1883. Four months later a crane knocked a Leith man named Dalgleish twenty feet off a cofferdam, cracking his ribs and leaving him with internal injuries.

Most injuries and deaths occurred when Briggers fell from heights or objects fell on them. One of many instances of the former occurred in June 1886 when a foreman mason named Robert Chalmers fell eighty feet from the steel girders onto rocks. One instance of the latter occurred in May 1887 when a piece of metal landed on Alexander Steel's head, killing him instantly.

There were some particularly nasty incidents, such as that of February 1885 when a staging broke, throwing twelve men into the Forth. There was worse the following month when a caisson gave way with seven Briggers inside. Rescuers saved five, and the two men who drowned were both married. These are only representative samples of literally thousands of accidents. Sufficient to say that there were so many deaths and injuries that men left the bridge to work elsewhere and officials asked hard questions of those in charge of the works.

Although the majority of Briggers were in their twenties and thirties, ages varied between fourteen and sixty. One unfortunate rivet boy, fourteen-year-old John Evans, fell from a height in January 1887 and received the doctors called 'serious concussion of the brain'. Three months later Thomas Birrell of Inverkeithing slipped and fell a hundred and fifty feet to his death. He was sixty years old.

There were other worries, including a shortage of water that saw a dispute between the North Queensferry works and nearby Aberdour, but which the authorities resolved when they built huge water tanks and imported water by sea.

In addition to the deaths and injuries, strikes for pay rises temporarily delayed the work. These included the outdoor workmen demanding an extra penny an hour in 1887, a strike in September 1888 about overtime payments and riveters striking in March 1888. The riveters hoped for a pay rise from seventeen shilling per hundred rivets to a pound per hundred rivets.

Disease struck in 1887 with scarlet fever hitting the workmen at South Queensferry, while the frequent Forth storms forced work to stop on a number of occasions including March and November 1888.

The numbers employed on the bridge varied. By June 1884 there were around 1200 workers, many living in bothies, with a canteen at South Queensferry. That number rose to 2,200 by April 1885 and included a number of Italians. In February 1887 over 3000 men worked on the bridge. Many of the men were labourers; some were brought in each day on special trains from Edinburgh; around a third were Irish. Some casual labourers worked for a few days before drifting on, and others were more used to the shipbuilding yards but took on bridge-building when that industry experienced a slump. At the height of construction as many as 4,600 men were employed, with the death toll variously recorded at between 57 and 73.

Naturally, many of the Briggers had fascinating and sometimes tragic tales to tell. For example, when eighteen-year-old William Duffy of South Queensferry died after plunging onto the rocks of Inchgarvie he left behind his recently widowed mother. There was also John Markey, a ship rigger. He was working such long hours at the bridge that at times he was unable to look after his young son. In June 1888 he gave him into the temporary care of Miss Stirling's Edinburgh Children's Aid and Refuge. Emma Maitland Stirling was half Scottish, half American and well known as the keeper of children's charity homes. She also sent some orphans to Nova Scotia for a better life.

When Markey's situation improved, he popped along to the amiable Miss Stirling to get his boy back. However, Miss Stirling told him that she had sent his son to Canada. Markey contacted the police, and they eventually brought the boy back to Scotland, with a court order specifying he should be at home. There were worrying times there for the unfortunate father but only good intentions by Miss Stirling.

Naturally, the construction of what was one of the engineering wonders of the age created a great deal of interest, and as well as boat trips from local ports, national and international dignitaries arrived to marvel at the bridge. August 1884 was a busy month for VIPs, with the King of Sweden and Norway visiting as well as the Prince and Princess of Wales and William Gladstone, the Prime Minister. The Shah of Persia and the Duke of Cambridge were also visitors to this international attraction. The Shah was apparently particularly impressed by the electric lighting that had been installed for the works as early as 1883. Prince Henry of Battenberg waited until 1888 before he graced the Forth with his presence.

It was not all work and drinking for the briggers. In 1883 the Roman Catholics planned to erect a church at South Queensferry and arranged a public subscription to raise the money while in August 1884 a train carried four hundred of the workers on a Saturday outing to Dundee. The briggers also attended the North Queensferry fair in August 1886, but when the drink flowed, fists flew, and the police had to break up the trouble. There were other kinds of games as well, with one Brigger, Joe Murphy, also playing football for Harp of Dundee; he may have watched the annual regatta around the growing bridges. There was also a prize fight with bare fists between an Englishman named Frank Snowden, and a Skyeman called Thomas McRae. The Englishman, far taller and heavier, won a bloody victory.

Finally, after years of speculation and graft, scores of deaths and far too many injuries, on Thursday 7th November 1889 the girder that linked the Inchgarvie cantilever to that of North Queensferry was carefully clicked in place; the final rivet slotted in place and the Forth was bridged. The following month the Bridge was ready for traffic. On 21 January 1890, with memories of the Tay

Bridge disaster still haunting Scottish travellers and engineers, the Bridge was tested for safety. Three railway engines, each towing fifty carriages with a combined weight of nine hundred tons, started at opposite ends of the bridge and rolled past each other. The Bridge held without a quiver. No passenger or goods train would carry such a weight: the North British were ensuring their bridge was safe.

With the Bridge passing all the necessary tests, the chairmen were the first to officially cross the entire length in February 1890 and then on the 4th of March 1890 vast crowds witnessed the Prince of Wales open the Bridge. No doubt the people of Queensferry would have breathed a deep sigh of relief when the Briggers departed, and they could reclaim their streets again.

Despite its fame and impressive presence, not everybody approved of this new example of engineering. In January 1890 George Washburn Smalley, the London correspondent of the New York Tribune wrote that 'it is not merely that the structure is hideous in itself, but that it totally ruins some of the most beautiful scenery in Scotland or the world.' He suggested that the designers and chief engineer should be hanged from one of the cantilevers.

With the Victorian love of statistics, every newspaper and journal carried details of the bridge, from the number of rivets used to its length, breadth and height. To summarise that, the Forth Bridge is over eight thousand feet long, three hundred and sixty-one feet high and holds the railway, so trains cross at one hundred and fifty-eight feet above the deep blue waters of the Forth. Over fifty thousand tons of steel were used, with millions of rivets, but the result is an engineer's dream of construction.

Since its opening the Bridge has operated day in, day out in all the weather the Forth has thrown at it; it witnessed a German submarine in the First World War and in November 1914 an anti-submarine net hung below the bridge. It survived that war, was a mute witness to Germany's first air raid on Britain in the Second World War and has since been used in films and documentaries. It was an isolated icon for over seventy years until in 1964 a companion joined it in crossing the Forth.

Crossing the Scotswater

If rail travel was the buzz word in the nineteenth century, the invention of the internal combustion engine revolutionised travel in the twentieth and, so far, in the twenty first.

While the Forth Bridge is a typical creation of the Victorian age, the Forth Road Bridge is a natural progression of a far older tradition. Although no doubt there were always some boatmen willing to take people across the Scotswater, it was not until the penchant for pilgrimage became popular in the high Middle Ages that an official ferry arrived. Pilgrims travelled all over Europe to visit holy shrines and worship their favourite saints. Scotland had surprisingly many of such sacred places, including Iona, Tain and St Andrews, which was possibly the most visited. To reach St Andrews from the south, pilgrims found it easier to cross the Forth than take the long detour to Perth or even Stirling for a bridge.

Queen Margaret, the Hungarian born wife of King Malcolm III, was said to have started the ferry between the two ports that bear her name, although the embarkation point in Fife shore carried the name of North Ferry for centuries. In 1130 King David 1, who increased the Norman influence in Scotland, improved the service around 1130. In time the holy men of Dunfermline could get free passage on the ferry, which the paying passengers may have resented, but rank has always had privilege, usually undeserved and granted to them by themselves. Nothing changes there.

The ferry saw some interesting personalities, from Mary Stuart, the queen who tore Scotland apart in the sixteenth- century, to rogues, outlaws and tens of thousands of ordinary decent people. One of the most unfortunate passages was that of King Alexander III, who crossed to Fife in wild weather to visit see his young wife, survived the voyage and promptly fell over a cliff, which led to Edward Plantagenet's spurious claims to be 'Overlord' of Scotland and the subsequent Wars of Independence.

In 1810 trustees took over the ferry, and in 1821 steam arrived. However, in time it became evident that the ferry was an inconvenient bottleneck in the smooth flow of traffic from the south to north or vice versa and from as early as the 'improving' eighteenth century there were ideas for a bridge over the Forth. Unfortunately, the technological and engineering skills of the period did not match the dreams, and a bridge remained merely a vision.

As traffic increased with the population, trade and transport expansion of the nineteenth century, more plans for a bridge were proposed and promptly discarded as the reality of the windy, often vicious passage hit home. No doubt the ferry operators were happy with the current situation: it kept them employed. Nevertheless, by the 1920s it was evident that motor vehicles were here to stay. In 1923 an Edinburgh journalist and visionary named James Inglis Kerr held a meeting at the Hawes Inn at South Queensferry and presented his ideas for a road bridge.

Serious work began to find the best spot and the most suitable approach. There was also some talk of a tunnel, but the economic realities of the depression, followed by the monstrosity of the Second World War held back progress; traffic continued to rely on the ferries. It was not until 1958, thirteen years after the end of the Second World War that a consortium of engineering companies formed for the mammoth project. The site chosen was nearly exactly where Queen Margaret and the pilgrims had crossed the Forth all these centuries ago: the old folk knew their Forth.

The Forth Road Bridge was to be a suspension bridge, the first of its kind in Europe. Engineers learned the techniques at special classes in South Queensferry, and then applied them over the Forth. Now a new generation of bridge builders worked high above the historic waters of the firth, with the bright blue bowl of the sky as an infinite ceiling and the eternal Forth wind plucking at them. As with the rail bridge, the road bridge was built in two halves: north and south stretching across to grasp hands above the centre of the Forth and the engineers proved their skill when both halves met with a difference of only one inch.

In common with the Forth Bridge, the Forth Road Bridge was – and is – an engineering marvel. At the time of its completion, it was the fourth longest suspension bridge in the world and the longest in Europe. It is over a mile and a half long; towers 512 feet above the Forth and took 39,000 tons of steel to construct, costing over nineteen million pounds. Again in common with the Forth Bridge, there was a human price to pay, with seven men killed in its construction.

On 4 September 1964 Queen Elizabeth opened the Forth Road Bridge in front of an estimated sixty-six thousand people. In the meantime the ferries had enjoyed their last run; their crews would have mixed feelings, but progress always has losers as well as winners. The Forth added suitable drama by providing a mist that obscured vision, a reminder of the difficulties faced by engineers and seamen alike. The Royal Navy fired a 21-gun salute and the Queen's highway now extended in an unbroken tarmac ribbon from Edinburgh to Perth and all points north. The Forth Road Bridge may have been one of the longest suspension bridges in the world, but it was not free, as traffic had to pay a toll to cross. The Scottish government removed that toll in 2008 and traffic now passes unhindered. The toll had paid for the initial cost of the bridge by December 1993, but the constant flow of revenue helped pay for maintenance. The bridge has also been strengthened to cope with the greatly increased volume and weight of traffic.

In the winter of 2015, an inspection found major structural faults, and the bridge and it was closed, first to all traffic and then to heavy goods vehicles. There was a political furore, with the opposition blaming the Scottish government for inadequate maintenance, but the real losers were the motorists. Only when the bridge was no longer in use was it properly appreciated, yet it must have been deeply frustrating to have a detour of scores of miles when a third Forth bridge was already under construction.

As early as the 1990s ideas for a third bridge were aired, and by 2000 it was apparent that traffic had multiplied to such an extent that a third bridge was needed to cope with the strain. The Scottish Government wasted no time in acting: consultation work began immediately.

Queensferry Crossing

The queen gave her assent to the proposed new bridge and again workers clustered to both sides of the Forth. The public was invited to suggest a suitable name and a short list created. There was a public vote over the three favourite names: Caledonia Crossing, Queensferry Crossing and St Margaret's Crossing, and Queensferry won. The legacy of Queen Margaret continued after nearly a thousand years.

Naturally, there were objections to this major new work, mainly from environmental groups who thought a new bridge was not essential, despite the increase in traffic and discovered structural weaknesses with the existing Forth Road Bridge. Once again there were proposals for a tunnel that would have been more environmentally friendly, but would also cost more and take longer to create.

At the time of writing [2017], the bridge is still taking shape and is expected to open in the summer of this year. The current idea is for the original Forth Road Bridge to carry public transport, bicycles and foot passengers, with the Queensferry Crossing with carrying all the rest. When complete, the new bridge will be the longest cable stayed three-tower bridge in the world, with an elegance that may even outmatch the Forth Road Bridge. It will be 2.7 kilometres, or 1.7 miles long, with the cables that cross mid-span ensuring extra strength.

Crossing the Forth, once an ordeal, should be a lot easier.

Epilogue

There have always been boats and ships on the Firth of Forth and probably always will be. The skills of seamanship have changed, but the basic requirements have not. It is true that satellite navigation and radio has removed much of the loneliness of seagoing, but the sea is still a dangerous place, and the future Forth mariner will, no doubt, be as pragmatic and practical as ever.

The Firth of Forth is as important to the nation today as it always has been, and while there are no longer English pirates or German submarines waiting to ambush the unwary, the sudden squalls can still be vicious, and the islands are as beautiful as ever. Bridges have tamed the waters and pleasure craft cram the marinas at Port Edgar, North Berwick and Anstruther, but fishermen still sail from the harbours, and trading vessels head for Leith. The Royal Navy keeps careful watch from Rosyth, families tread the beaches at Yellowcraigs and Pettycur, while seagulls and seals use the Forth for their own ends.

As always in the Forth, history will write itself. The future will have its own stories, as does the past and the people who live on either side of the firth will be as kindly and hard working as ever. After all, it is the people who matter.

Malcolm Archibald 2017

Select Bibliography

Anson, Lord, *A Voyage Round the World 1740-1744* (London, 1911)

Anson, Peter, *Fishing Boats and Fisher Folk on the East Coast of Scotland* (London 1930, 1971)

Archibald, Malcolm, *Scottish Battles* (Edinburgh 1990)

Archibald, Malcolm, *Whalehunters* (Edinburgh, 2004)

Brodie, Ian, *Steamers of the Forth* (Plymouth 1976)

Brown, Hamish, *The Fife Coast* (Edinburgh 1994)

Caledonian Mercury

De La Varende, Jean, *Cherish the Sea* (London, 1955)

Dundee Courier

Edinburgh Evening News

Falkus, Hugh, *Master of Cape Horn* (London, 1982)

Fraser, Duncan, *Discovering East Scotland* (Montrose 1973)

Hough, Richard, *The Great War at Sea 1914-1918* (Oxford 1983)

Kemp, Peter (Editor), *The Oxford Companion to Ships and the Sea*, (Oxford 1976)

Kohli, Marjory, *The Golder Bridge: Young Immigrants to Canada 1833-1939* (Toronto, Ontario, 2003)

Lubbock, Basil, *The China Clippers* (Glasgow 1914)

Lubbock, Basil, *The Arctic Whalers* (Glasgow 1937, 1978)

Mackay, S, The Forth Bridge, (Edinburgh, 1990)

McGowran, Tom, *Newhaven-on-Forth: Port of Grace* (Edinburgh 1985)

McKean, Charles, *Battle for the North* (London 2006)

Miller, James, *Salt in the Blood*, (Edinburgh 1999)

Mowat, Sue, *The Port of Leith* (Edinburgh nd)

Smout, T. C. *A History of the Scottish People 1560-1830* (London 1969)

The Scots Magazine
The Scotsman
Winston, Alexander, *Pirates & Privateers* (London, 1972)

About the Author

Born and raised in Edinburgh, the sternly-romantic capital of Scotland, I grew up with a father and other male relatives imbued with the military, a Jacobite grandmother who collected books and ran her own business and a grandfather from the legend-crammed island of Arran. With such varied geographical and emotional influences, it was natural that I should write.

Edinburgh's Old Town is packed with stories and legends, ghosts and murders. I spent a great deal of my childhood walking the dark streets and exploring the hidden closes and wynds. In Arran, I wandered the shrouded hills where druids, heroes, smugglers and the spirits of ancient warriors abound, mixed with great herds of deer and the rising call of eagles through the mist.

Work followed with many jobs that took me to an intimate knowledge of the Border hill farms to Edinburgh's financial sector and other occupations that are best forgotten. In between, I met my wife. Engaged within five weeks we married the following year, and that was the best decision of my life, bar none.

At 40 the University of Dundee took me under their friendly wing for four of the best years I have ever experienced. I emerged with a degree in history, and I wrote. Always I wrote.

Malcolm Archibald

Lightning Source UK Ltd.
Milton Keynes UK
UKHW011842081220
374864UK00008B/449/J